A FIRST CLASS JOB !

A FIRST CLASS JOB !

The story of
FRANK MURPHY

radio pioneer, furniture designer
and industrial idealist

Published by the author
JOAN LONG
Sheringham   Norfolk

Copyright Joan Long 1985

All rights reserved. No part of this publication may be reproduced in any form or by any means without the prior permission of Joan Long, Weybourne Road, Sheringham, Norfolk

Printed in Great Britain by
Roger Pritchard   New Costessey   Norwich

To

CHARLES RUPERT CASSON

a true friend
who has never ceased to regard "F.M."
with admiration and affection—

In love and gratitude

Joan Long

"F.M."

ACKNOWLEDGMENTS

In venturing to write this biographical sketch of Frank Murphy I have had immense help and encouragement from a very great number of people who knew and worked with him. Foremost among these is C. R. Casson, who was able to recall so vividly conversations and scenes of more than fifty years ago. I must also give especial thanks to those "Murphy-Mad" types, George Denton, Clifford Stephenson, Tom Rowney, Len Searle, George Berry, Frank Thompson, Louis Driscoll, Douglas Boyd, Doris Hulks, Mrs. E. Trimble, John Pank and the late Alec Vallance for their reminiscences and their hospitality. To Ted Bromley I owe the loan of ten weighty volumes of the *Murphy News*—an invaluable source of material. Jack Hum not only encouraged me in my task, but gave me much expert help and editorial advice, and he and his wife Grace have always welcomed my visits to report progress. Mr. Angus Crichton-Miller and John Westlake, formerly Managing Director and Publicity Officer for Rank Radio International, were also most helpful. I spent a happy day in the BBC Written Archives Department at Caversham, and owe them permission for the right to reproduce some of the old *Radio Times* advertisements. Michael Russell kindly gave permission for extracts from the late Sir Gordon Russell's book, "Designer's Trade", and the late Beverley Nichols kindly allowed me to reprint an extract from an article of his in the *Radio Times*. Particular help came from Douglas Boyd and Max Broughton when I was preparing the chapter on the Frank Murphy furniture, which I hope may now receive some overdue recognition.

Finally, I must thank my former secretary, Diane Young, not only for typing the manuscript, but also for returning each chapter with a humorous comment—in verse!

J.L.

CONTENTS

FOREWORD		xiii
CHAPTER 1	Unusual Principles of a Businessman	1
CHAPTER 2	Early History of a Mathematical Prodigy	8
CHAPTER 3	The Publicity Business	19
CHAPTER 4	The Beginning of Murphy Radio	26
CHAPTER 5	"Now We'll Tell Them!"—The Murphy Advertising	34
CHAPTER 6	VALUE FOR MONEY—Early Years of Murphy Radio	52
CHAPTER 7	"MURPHY MADNESS!"—The Dealers Story	73
CHAPTER 8	Spreading the Gospel—The *Murphy News*	93
CHAPTER 9	A New Conception of Business	124
CHAPTER 10	Good Furniture for All	133
CHAPTER 11	Marketing Plans for Murphy Furniture	158
CHAPTER 12	The Wilderness Years	168
CONCLUSION		191
APPENDICES		
APPENDIX 1	The Murphy Family Tree	193
APPENDIX 2	Articles and Booklets by Frank Murphy	194
APPENDIX 3	Employment and Working Conditions at Murphy Radio, 1932-36	196
INDEX		201

LIST OF ILLUSTRATIONS

Fig 1.   Frank Murphy as a young man (1908)
Fig 2.   Frank Murphy in R.F.C. uniform (1917)
Fig 3.   The Murphy family and friends (1920) (F.M. at centre back)
Fig 4.   Hilda Murphy with Kenneth and Joan (1923)
Fig 5.   Frank, Hilda, Maurice and Joan on top of Snowdon (1935)
Fig 6.   Teatime music in camp with a B4 portable (1929)
Fig 7.   Relaxing after a Girl Guide camp (Hilda and Dorothy)
Fig 8.   Test bench (B4's in foreground)
Fig 9.   The Welwyn Stores in the early '30's
Fig 10. The Cherry Tree Restaurant, Welwyn Garden City
Fig 11. The house at 25 Brockswood Lane
Fig 12. Murphy Radio's new factory (1931)
Fig 13. Ludwick Corner in 1930
Fig 14. Ludwick Corner in 1981
Fig 15. Murphy Radio's first assembly line
Fig 16. Press Shop: Murphy Radio
Fig 17. The A26 Radiogram (1934)
Fig 18. "Dignified and Beautiful" the A24C
Fig 19. The A28 Radiogram (1935)
Fig 20. The first Murphy television set (1937)
Fig 21. The First Murphy factory
Fig 22. Progress from 1930 to 1934
Fig 23. Frank Murphy with two of his dealers - F. Piggott of Kingston-on-Thames, and James Thornes of Dewsbury
Fig 24. Dealers' Publicity (1)
Fig 25. Dealers' Publicity (2)
Fig 26. Dealers' Publicity (3)
Fig 27. New Technique in the B27 cabinet
Fig 28. The 'Chinese Oval' of the A30 loudspeaker
Fig 29. The A34 (1937)
Fig 30. The B31 (1937)
Fig 31. Inside cover of Frank Murphy Ltd. furniture catalogue
Fig 32. Kenneth Murphy (1938)
Fig 33. Audrey Murphy (1932)
Fig 34. The staff of Frank Murphy Ltd at Ludwick Corner (1939)

Fig 35. A. R. Turner (1938)
Fig 36. The cabinet-making shop at Ludwick Corner
Fig 37. Old Windsor Chair
Fig 38. Types of joint used in Murphy tables and chairs
Fig 39. The Murphy dining chair
Fig 40. The fixed dining table
Fig 41. The 5-foot sideboard (1)
Fig 42. The 5-foot sideboard (2)
Fig 43. The 4-foot sideboard, drawleaf table and dining chairs
Fig 44. Test for scratching of table top surface
Fig 45. Racking test on chair
Fig 46. Model chair in clay used to find correct shape and angle for back
Fig 47. Mould in modelling clay to ascertain correct seat shape when bodily position is unchanged
Fig 48. Mould in sand based on previous clay mould, but allowing for change in bodily position
Fig.49. Underneath of fixed table
Fig 50. Mechanism of drawleaf table
Fig 51. Frank with Maurice in Toronto (1953)
Fig 52. The Murphy Phonograph made in Canada (1948)
Fig 53. Hilda Murphy aged 85 with great-grand-daughter Deborah Long (1972)
Fig 54. Hilda Murphy aged 90 receiving a Murphy TV set
Fig. 55. Last portrait of Frank Murphy, taken in Canada
Fig. 56. Poem by Frank Townshend (sent by F.M. as a Christmas card)
Page 36 MAKING WIRELESS SIMPLE"
Page 38 "THAT'S FINE—NOW GET ME ANOTHER STATION, JOAN..."
Page 44 WHAT A LIFE!
Page 46 UNEMPLOYMENT!
Page 50 WHAT IS CHEAP?!
Page 51 1ST PRIZE FOR THE NEW MURPHIES!
Page 64 "THIS IS MY NEW SET"(the A3)
Page 70 "THE LONGER YOU LOOK AT IT, THE MORE YOU LIKE IT!" (The A24)
Page 71 A COMPLIMENT IMPLIED—-AND ACCEPTED!
Page 74 MURPHY MADNESS!
Page 100 ONE DEALER PER SHOPPING CENTRE

FOREWORD

To say that the times find the man is a truism. To say that a man becomes a legend in his own lifetime is to risk an over-worked cliché. Nevertheless, history is full of examples that bear out these claims. One such man was Henry Ford. Another, in the early formative years of the British radio industry, was Frank Murphy, the subject of this book.

Like Ford, Frank Murphy was an engineer who saw that what people wanted was value for money in something they could use every day, a product they could be proud to make, to sell or to own. Murphy's insistence on a totally honest approach in business dealings by all who worked for him, and his emphasis on service to the consumer as a greater priority than profit-making may have derived from Ford. But what was unique was his ability to communicate his ideals and to fire with enthusiasm—or "Murphy-Madness", as it came to be called, many hundreds of diverse followers.

From a successful career in advertising he turned to radio, and from radio to domestic furniture, and ultimately to a critical look at the whole field of industrial relations within the capitalist system. Only now, thirty years after his death, is it possible to see how far-reaching were many of his ideas, and how much in advance of his time.

It was fortunate for Frank Murphy that wireless broadcasting was one of the two technological miracles to emerge from World War I (the other being aviation), and that by his training, his experience and his zest for research he had built up a greater knowledge of radio than most of his contemporaries.

In 1922 the British Broadcasting Company (later the British Broadcasting Corporation) began its first transmissions, offering a single service from a chain of small low-powered stations. Within a very few years the BBC had developed a service of nationwide alternative programmes, and this called for the massive provision of valve receiving-sets throughout Britain. The rudimentary crystal-detector receivers ("cat's whiskers") of the early enthusiasts steadily gave way to simpler, more reliable models for family use. Then, as the manufacturers developed more sophisticated receivers, particularly those employing a super-heterodyne circuit, another need arose. This was for a generation of technically-trained radio dealers, to replace the old "hit or miss" wireless suppliers of the early days.

As the time came for mass production of high quality receiving sets, Frank Murphy, the wireless expert and natural entrepreneur, was there. In 1929 he found in Welwyn Garden City an ideal place for living and working in. Completely unknown and with the barest minimum of capital, he launched his new company, Murphy Radio Limited, convinced that his sets were so good and so reliable that the public would readily buy them. He was right. In less than seven years the Company had built up its turnover to over a million pounds per annum—a remarkable achievement in the Depression Years of the Thirties.

In 1938 Frank Murphy began a new venture—an attempt to bring high quality furniture of outstanding design within the reach of the average British family. Because of the onset of World War II, the business, like many others at that critical time, could not continue. It is probably true to state that few people today are aware that Murphy furniture pre-dated and outclassed the much better-known "Utility" furniture by nearly a decade. The time is ripe for a reassessment of this phase of Frank Murphy's activities.

Liked all outstanding men, Frank Murphy was no easy man to live with. He had the faults that go with burning enthusiasm and a highly individual temperament. But he had faith in his ideals, a natural love of his fellow men, warm affection for his family and his many friends, and a tenacious desire to "get to the heart of things". Small wonder he became a legend in his time.

JACK HUM
(Editor of the *Murphy News*, 1946-1966)

Chapter One

UNUSUAL PRINCIPLES OF A BUSINESSMAN

"The truth shall make you free." *John 8, 23*

"Good heavens, what on earth's that?" Mrs. Brown and her friend were on their homeword journey after their usual Saturday shopping trip in 1934 in North London. Peering out of the window, they could see a tradesman's motorcycle combination going past them, driven apparently by a man with a gigantic head. In his smiling jaws an equally large pipe was clenched. The back and front of the sidecar were inscribed with the eye-catching message, "Buy Your Murphy Now!"

"Oh, of course,—it's our local Murphy Dealer—that's one of Frank Murphy's carnival heads he's wearing."

"Frank Murphy? Who's he?"

"Oh, you know. The Murphy Radio man that advertises in the Radio Times. My brother Sid bought one of their sets last month—lovely tone it's got, and it wasn't expensive, either."

Sid and his wife, like countless other British families in the nineteen-thirties, led hard-working and rather humdrum lives. Sid's weekly wage as a railway porter provided little scope for luxuries, but the family now had a much-prized source of information and entertainment—their radio set. Through it they heard the familiar voices of John Watt, Frank Phillips and Alvar Liddell, giving them the national news and sports commentaries and results. They chuckled at Stainless Stephen, Stanley Holloway and Gracie Fields. They relaxed as they heard the familiar strains of Jack Hilton's band, and the sophisticated sound of Carroll Gibbons at the Savoy (a place they were never likely to frequent). Probably they weren't interested in classical music or Greek tragedy or grand opera, but these were all becoming available to a new mass audience through the courtesy of the British Broadcasting Corporation. A whole world of new experience was opening to them—thanks to the arrival of the mass-produced radio receiver which, as Frank Murphy had now triumphantly demonstrated, could be simple to operate, reliable, and first-class value for money.

How did this brash newcomer to the radio industry, starting out in 1929 with incredibly small resources, manage to capture a sizeable part of the market and to become a household name into the bargain? Some of Frank Murphy's rivals, baffled at his success, tried to copy features of the Murphy sets and even the Murphy style of very personal advertisement. It did not affect the company's progress. The reason was that Murphy Radio was based on a set of unusual principles, refreshingly simple and rigorously applied.

The first principle was *"Value for Money"*. Having had a scientific training, Frank Murphy knew that any task must be undertaken to the very best of his ability if it was to give lasting satisfaction to himself and others. From early socialist and religious conviction he believed in "the greatest good for the greatest number". Making a fortune by selling elaborate expensive toys to the rich did not interest him—what he wanted was to provide the vast majority of British families with a good, simple and reliable wireless set at a price they could afford to pay.

The price of the first Murphy table model at seventeen guineas (£17 17s. 0d.) was certainly cheaper than its rivals; but its main advantage lay in the simplicity of its controls, so that even a child could operate it. Moreover, each component used in the Murphy set had been put through the most rigorous tests the company's laboratory staff could devise, and nothing below these standards was allowed to pass. It was thus possible to guarantee reliability to a degree previously unheard of in a mass-produced set. Faults, if they did occur, were speedily investigated and put right at little or no cost to the customer. Pride in their workmanship was a spirit that spread right through the Murphy works, and fired the enthusiasm of Murphy representatives, Murphy dealers, and even Murphy owners. Later, we shall see some of the ways in which this unique enthusiasm, or "Murphy Madness", as it came to be dubbed, was manifested.

Buyers of Murphy sets were certainly getting value for their money; but Frank Murphy felt that this was not enough. Very few people at this time knew how a wireless set worked, or could operate it efficiently. They needed expert help, and also protection from exploitation by unscrupulous salesmen. Frank Murphy's answer was to ensure that only specially selected dealers, whom he or his trained representatives could vouch for, were allowed to sell Murphy sets.

This Limited Dealer scheme was unique. Up till this time the general pattern in the radio trade was for manufacturers to sell to wholesalers, who in turn sold to retailers, who then sold to the general public, and the customers naturally had to pay a price that covered the profits demanded by the intermediaries. Frank Murphy argued the wholesalers were not giving a service to the public, and stated that his company would deal directly with the retailers—and then only a selected few of those. The wholesalers were naturally furious, and one of them remarked acidly, "He's bound to fail with this mad scheme. You'll see—we'll run him out of business in a matter of weeks."

Fortunately, the prophecy failed to come true. Frank Murphy was able to announce in October 1930 that fifty-nine dealers had joined his scheme, and two weeks later in a *Radio Times* advertisement he claimed:

"YOU DON'T KNOW ME YET—but you soon will! That is what I said to the wireless trade only four months ago. Now the whole wireless industry has heard of us and what we are doing. Nearly two hundred first-class dealers all over the country are now selling Murphy sets and helping me to make wireless simple."

To emphasise that these dealers were appointed specifically to serve the public, the company supplied them with showcards and banners in bold colours, saying "YOUR MURPHY DEALER".

The word "YOUR" was shrewdly chosen. Not only did it reflect the personal relationship between the Murphy dealers and their customers; it also embodied Frank Murphy's second principle—*Mutual Co-operation in the Service of Society*. (Though he usually expressed it in more homely terms, such as, "We've got to pull together as a team.") Such an idea might be expected to issue from a philosopher or a minister of religion, but it was odd to discover it in a hard-headed manufacturer.

In the tough world of commerce most directors would privately concede that they were basically determined by the profit motive, with little real concern for either retailers or the public. Similarly, many radio retailers prior to the Murphy era thought of their customers merely as items on their cash registers, rather than as potential friends. As an article in the trade press put it: "We (i.e. the radio trade) have a common enemy—the buying public. . . . How are we going to combine against the man-in-the-street?" But Frank Murphy, on the other hand, made it clear that "MURPHY DEALERS ARE PEOPLE YOU CAN TRUST."

Just as the public was encouraged to trust the special Murphy dealers, so Frank Murphy showed that he likewise was prepared to trust them. The dealers were amazed by the company's willingness to listen to their problems, to confide in them its plans for new models, changes in prices, advertising campaigns, etc., and above all, to involve them in the thinking behind the Murphy policy. In every region in the United Kingdom dealers were regularly invited to a first-class hotel for a free meal, followed by absolutely candid discussion on company policy, with experts there from Murphy Radio to answer all their questions. As one dealer remarked, "What other manufacturer cares two hoots about us?"

This was not all. In the Murphy factories, every single person from the newest junior on the assembly line to the most senior executive was encouraged to believe that they too could contribute to the company and its ideal of service to the public. They felt that what they did mattered, and the atmosphere of comradeship in the place was as much appreciated as the good rates of pay. Frank Murphy made a point of walking round each day so as to get to know them all personally. He had attracted an outstanding team of young men to his laboratory, whose technical advances ensured the company's leading opposition in the industry. According to J. D. A. Boyd, one of the chief technical designers, the "Guv'nor" would frequently stop beside a man and enquire bluntly what he was doing and why. If the man could give an intelligent answer, all well and good; but woe betide him if he was merely proceeding by rule of thumb! On the other hand, praise was always immediate and unstinted if it was deserved—"First-rate! That's a first-class job!" would be a typical comment. Visits by dealers to the Murphy Radio factories at Welwyn Garden City were always welcomed, and the dealers in turn could contact specialists in the different departments for help and advice.

Frank Murphy's third principle, linking inevitably with the first two, was the necessity for *honesty and trust in all human relationships.* This basic attitude to life he defined as *"Integrity"*. It was a key-word for him, and it cropped up frequently in his speeches at dealers' meetings, in his daily conversations, and in the Murphy advertising as well. "Integrity", "Fair dealing", "Fifty-fifty", "An Englishman's word is his bond"—they were all, as Frank Murphy saw it, expressing the same concept.

It was in sharp contrast to the prevailing business ethic of that time, which could be summed up in the phrase, "Let the buyer beware!" While manufacturers tried to make as much profit as they could out of a gullible public, retailers did the same by selling shoddy goods when they could get away with it, and shrewd customers used backdoor methods to get what they wanted at less than the market price. Complete and unqualified honesty was a rarity. If a piece of sharp practice was detected, the comment generally was, "Business is business, I suppose." Not that most people were actively dishonest, but it was generally thought that if a man's livelihood and the welfare of his family depended on his business acumen, he could scarcely be blamed for trading on the other fellow's ignorance. This state of affairs was particularly true in the radio industry in the nineteen thirties, when electrical knowledge was a technical mystery to all but a few, and the ordinary buyer could easily fall a victim to the wiles of high-pressure salesmen.

Now here was a radio manufacturer saying to the public,

"You can trust me, and you can trust my dealers. In return, I expect you to pay the agreed price for my set, which is as low as we can make it in our present state of knowledge."

To his dealers he said,

"If you want to remain a Murphy Dealer, you have got to show that you are genuinely improving your knowledge and technique in servicing radio sets, and not merely going out to achieve more and more sales. You must examine how you run your business, how you train your assistants, and how you can best look after the needs of your customers. In return we will tell you at the beginning of each year which models we will be bringing out and what our prices will be, so that you can plan ahead. Although you will be getting a smaller discount than other (non-Murphy) dealers, and we shall expect you to pay cash within seven days of receiving any sets you order, you will have the security of being the only dealer able to sell Murphy sets within your area".

Such simple directness was not mere naiveté. Frank Murphy well knew that clever and dishonest men can make fortunes at the expense of others, but he firmly believed that in the long run, honesty does pay, and is in fact the only viable policy for human society.

The fourth principle which underlay Frank Murphy's approach to life and to business was the passionate desire *to solve problems, and to get at the truth*. Like Henry Ford, he saw that when you can define a problem, you are more than halfway towards solving it. Like Henry Ford again, he realised that many seemingly complex problems are not solved by adding further refinements, but by cutting out unnecessary processes or theories, and "making it simple". Right from the start he was to adopt the phrase "Making Wireless Simple" as the Murphy trademark, and it appeared as a banner headline in all the early advertisements. He even wrote a booklet with the same title, which was sent out free to any member of the public who wrote in for it.

"Well, what's the problem?" was Frank Murphy's frequent greeting to harassed dealers and nervous new representatives. Indeed, he was likely to say it to anyone he met. Perhaps up to that moment, they were unaware that they had a problem. But very soon, after ten minutes' acute questioning, the victims suddenly saw not merely the problem, but its probable solution— all they had to do was to follow Frank Murphy's advice, and the outlook would be much brighter! This is, of course, an over-simplification, but there is no doubt that most people went away distinctly heartened after a talk with him, even though the initial probing session could be painful. It may explain why the famous dealers' meetings had such a magnetic appeal—hundreds turned up to them, and the numbers rose steadily year by year—because however mutinous the dealers might be about his latest proposals, they knew that he was putting them forward for discussion, not as a directive, and he genuinely wanted the truth to emerge.

Each stage of Murphy Radio's development illustrates this "problem-solving" technique. The problem of serving the public was met by designing a set that was utterly simple to operate and seldom, if ever, went wrong. The problem of protecting Murphy owners was met by the Limited Dealership Scheme, by appointing only men who could be trusted to be of real service. The problem of finding suitable cabinets for housing the radio sets led to the involvement of a leading furniture designer in the earliest stages of design. The problem of two-way communication between dealers and the staff of Murphy Radio was met by the founding of the *Murphy News*. The problem of servicing the dealers led first to advice on window display, then to advice on advertising, showrooms, service departments, accounting and management, and to training of dealers' staff by specialists from within the company.

Frank Murphy left Murphy Radio because he wanted to solve the problem of human relations in industry, and because he did not think it could be done within the framework of the traditional limited liability company. As we shall see, he attempted a solution with his "New Conception of Business". When World War II came, he had further problems to face:—his attitude to war and to pacifism; post-war planning controls restricting new businesses; the death of his elder son and the breakdown of personal relationships within his own family; and his own loss of status as a valued member of the business community. Whatever the cost, Frank Murphy continued to define the problem as he saw it, and to attempt a solution. He believed that "the truth shall make you free".

Chapter Two

EARLY HISTORY OF A MATHEMATICAL PRODIGY

"The Child is father of the Man." *Wordsworth*

Despite his surname, Frank Murphy was not an Irishman—though his grandfather had come from Dublin. Frank's parents were John James Murphy and Annie Leggo, who had met and married in Cornwall. John Murphy had trained as a teacher at St. Loyes' College, Exeter, and set up home in London when he obtained a teaching post there, and eventually in 1890 became Head Master of Holmes Road School in Kentish Town. He took his profession seriously and played an active part in his local branch of the National Union of Teachers. A well-respected member of the community, he joined the Volunteer Force, and was a Master of two Freemasonry Lodges. At home his wife held him up to the children as the head of the family, worthy of unquestioning respect; but in later life they came to suspect there might be serious flaws in their father's conduct, deriving from those very qualities of charm and bonhomie which made him so popular. His youngest daughter, Ethel, described the contradiction in this way;

"My father was a jolly, sociable person, with twinkling blue eyes, striking red hair and 'Kaiser Bill' moustache. He dressed well, he loved to dance, he had a good baritone voice, and he was a passable musician on both the piano and the organ. He loved to see his family dressed for church on Sundays, and took pride in them both as a group and as individuals. But he was careless about money, which worried our thrifty mother, and in his latter years he undoubtedly drank too much, and eventually died of Bright's disease brought on by his intemperance. Nevertheless, he was sincerely repentant for all the trouble he caused, and Mother loved him tenderly in spite of his faults."

Annie Murphy's life was certainly not an easy one. She bore eleven children, four of whom died in infancy, and it was a hard struggle to bring up the surviving four boys and three girls and keep up appearances as the wife of the local Headmaster. Frank described her as "a very gentle, patient woman who never got angry with us, however troublesome we were. The worst she ever threatened to do was to stick her darning needle in us (she was everlastingly darning our socks), but needless to say, she never actually did so". Frank and Ethel, the two youngest children, were born at Penrith Villa, 25 Wrottesley Road, Plumstead, in East London,—Frank on June 16th 1889, and Ethel in the following year. By the time Ethel was three years old the family had moved to 22 Gladsmuir Road, Upper Holloway, which was nearer to John Murphy's school in Kentish Town. She and four year old Frank were then started at Whittington School, although Frank quite often attended only in the afternoons, apparently because his doting mother felt he should not be forced to get up until he felt like it. On their way to school, the children were accompanied by the family dog, a large black retriever called Nero, who also called for them when it was time to go home.

Before long, however, there was another upheaval. At John Murphy's insistence, the family moved back once more to the Plumstead house which he had had built for them and where he fondly imagined they would all be more comfortable.

Unfortunately there were no private schools in Plumstead suitable for Ethel and Frank, and Annie did not think it fitting for a Headmaster's children to attend the local Board School. "What does it matter, for a few years?" was her easy-going husband's rejoinder. "When they are eleven, they can sit for a scholarship and travel a few miles further to a good private school." So indeed the children did, and like their elder brothers and sisters, they seem to have had little difficulty with school examinations, and generally came top of their classes. Frank, certainly, seems to have been a glutton for knowledge, and at a very tender age showed a special aptitude for mathematics. According to Ethel he was remarkable in showing no great enthusiasm to depart when the school day ended, and once home, he would setttle down to his favourite occupation. "Old Ponderous", as the family nicknamed him, was never happier than when working on some difficult equation in a quiet corner of the room, while the rest of the family were chattering away about the day's doings.

No one was greatly surprised, therefore, when Frank won a scholarship to East London College in the University of London. Even the family sat up, though, when he graduated with honours in electrical engineering at the early age of nineteen, and following this he was awarded a Mathematical Exhibition to St. John's College, Oxford.

It ws hoped that an Oxford degree would lead to entry into the Indian Civil Service, but this was not to be. After only a single term at Oxford, Frank had to abandon his course, for two reasons. The first was that he had not learnt Latin or Greek, then an essential entrance qualification for Oxford undergraduates; the second was that, due to his father's inroads on the household budget, the family could not afford to meet the remainder of his fees not covered by the Exhibition. Whether if circumstances had been different he might have ended as a university professor is an interesting speculation. Probably not; but he always retained his respect for academic achievement and was inordinately pleased when, a generation later, his daughter obtained a Double First at Cambridge.

Frank's mother understandably wanted him to enter the Civil Service, with its secure career and guaranteed pension prospects. The two branches which seemed most suitable were the Patent Office and the Post Office (in the research sections). The Patent Office entrance examination contained questions on chemistry, about which Frank knew virtually nothing, so he felt it would be safer to attempt the Post Office examination. But as he was still only nineteen and could not even apply for the Civil Service until he was twenty-one, he looked round for an interim occupation.

His first job was as a telephone assembly worker in the North Woolwich works of the Western Electric Company. For this he received a wage of 4½d per hour, or nineteen shillings and sixpence (£0.97½p) per week. This was later increased to the meagre sum of twenty-six shillings (1.30) per week. The hours of work were long and the work itself dull and repetitive. The workers had to be at the factory by 7.30 a.m. but little work was done until the foreman arrived at 9.00 a.m. By the end of a very long day Frank was thoroughly bored and frustrated, so every night he went back to East London College to do post-graduate study on a fascinating new development in electrical engineering-the cathode ray tube. This was the forerunner of today's television tube. Eventually the excessive hours took their toll, and Frank went down with a heavy bout of influenza. When he recovered, he was not sorry to find that his post at Western Electric had been filled.

His next job, he decided, must be an easier and better paid one. Fortunately, he succeeded in getting a post as an examiner with the International Correspondence Schools in London, at a salary of £3.00 per week. This was ideal, because the more flexible hours gave him time to study for the Post Office entrance examination. Within a year he had passed it triumphantly, coming out top of three hundred candidates. The next hurdle was the Interviewing Board. Frank had to face a large number of vociferous examiners, one of whom shot at him the final question: "Now, Mr. Murphy, why did you apply for a position in the Post Office?" Feeling that a strictly truthful answer—the lack of any questions on chemistry—might not be too well received, he replied diplomatically that he thought the work might be interesting!

His occupation settled, Frank was free to organise his leisure, in what remained of waking hours. His chief recreation at this time was hockey, at which he pronounced himself to be "not brilliant, but not too bad". His other leisure activities were tennis, cycling and walking. He became quite keen at tennis, in fact, winning not through excellent ground strokes, but by brilliantly wrong-footing his opponents—whereupon he usually laughed uproariously.

Basically, however, he remained a sober and serious young man, attending the local Baptist Chapel at Highgate Road three times on Sundays. He was even prepared to offer himself for missionary service. The Highgate Road Chapel attracted many young people to its services, partly because of its minister's fine preaching, and partly because the elders ran a flourishing weekly youth club. But while encouraging the social meeting of young men and women, the leaders taught or implied that sexual desire was wicked and must be sternly repressed by anyone aspiring to be a true Christian. This undoubtedly troubled Frank during his "religious" period, but having the ordinary instincts of a healthy young man, he finally decided that the life of an ascetic was not for him. On the contrary, he found himself very attracted to a young lady with deep brown eyes and long dark hair, who also attended chapel regularly.

Many years later he commented, "Going to chapel, and being normal, the normal thing happened to us. Boy met girl, and having sufficient money, we married at the age of twenty-two, and the marriage continued reasonably happily for thirty years. . . ."

This very terse account of how Frank Murphy met and eventually married Hilda Constance Howe gives very little idea of the immense importance this "reasonably happy marriage" was to have in their lives. Frank's brilliant intelligence, together with the messianic fervour of his temperament, caused him to alternate between the heights of exaltation and the depths of despair; today he would doubtless be termed a manic-depressive. Hilda provided the steady comforting presence of one who truly believed in his potential, while at the same time curbing his wilder excesses with a mixture of gentle humour and practicality. They "walked out" for several years, saving up for marriage like most conscientious young couples of their time. Their pleasures were simple. One of Hilda's abiding memories was of being met by Frank and sharing the results of her evening cookery class with him under the lamplight as they walked home. In 1912 they were married where they had originally met, at Highgate Road Chapel, and they set up home in rented accommodation at Ealing in West London. Their first son, Kenneth John Darby, was born on Derby Day, 5th June 1913.

Hilda Howe came, like Frank Murphy, from middle class parents, but unlike him, she was an only child. Born on 25th October 1889 at 6 Marlborough Road, Bowes Park in North London, her father was Charles Howe, the second son of a Hertfordshire gentleman farmer, whose early expectations of affluence never materialised. He married a young teacher, Edith Elizabeth Starr, whose background and personality were very different. Charles Howe was untrained for a profession or even a trade, and liked nothing better than a day's fishing, or a leisurely ramble by the river. Edith was made of sterner stuff, and was bitterly disappointed at Charles' lack of discipline and ambition. Worse still, he took to going to the races and lost what little reserve of money they had. Because Edith could not support him after the birth of little Hilda, he was reluctantly persuaded to try for a job as an agent with the Prudential Life Insurance Company. As one might expect, he was no salesman; he wanted nothing more than a quiet dreamy life in the country. When he took Hilda out for a walk to escape Edith's sharp tongue, they experienced rare moments of child-like happiness. Edith, meanwhile, was having a hard task to make ends meet and to keep up a respectable appearance before the neighbours in Kentish Town. She felt that it was a pity to let Hilda mix with "rough children" at the local Board School and she decided to teach her at home herself until Hilda was eight years old.

By this time the little girl was firmly imprinted with her mother's ideas on cleanliness and proper behaviour. Accordingly it was considered safe to send her to Burley Road School in Kentish Town, where she did very well. With her natural high spirits and sense of fun, she soon made friends with other girls in her class. Her special friend was Esther Marshall, another only child, who was later to attend the North London Collegiate School, and ultimately become Headmistress of Fulham Junior School.

For Hilda, on the other hand, there was no chance to become a teacher. Her parents could not afford to keep her at school beyond the age of fourteen, so she decided to train as a shorthand-typist, the only other career open to her (apart from working in a shop or in domestic service, which her mother certainly would not have countenanced). She took a course at the Empire Typewriter Company, and soon became the star pupil, achieving speeds of 200 words per minute in shorthand. This would certainly have guaranteed her an excellent secretarial post when the time came, but Hilda's passion for musical shows led her one day to organise a line of fellow pupils in a demonstration of high-kicking. In the middle of this enjoyable display, the Principal of the College walked in. Furious at this disruption of discipline, she announced that Hilda Howe would go forthwith to the very next vacancy notified to her establishment.

Consequently Hilda found herself banished to a dusty room in the offices of a firm of solicitors, Fraser and Christian, at the back of Lincoln's Inn in Central London. Here she spent hours faultlessly typing long incomprehensible legal documents for one elderly solicitor's clerk. She did, however, have two compensations. One was that during her brief lunch hour she was free to roam the streets of London, or sit in the quiet gardens of Lincoln's Inn. The other was that the crusty old clerk became unexpectedly human one day, when he discovered that Hilda had never read any novels by Charles Dickens. He at once encouraged her to read them, even lending her his copies. Soon she, too, was filled with his enthusiasm for the great arrray of Dickensian characters.

Hilda's former school-friend, Esther Marshall, had now made another friend at her new school; her name was Ethel Murphy. One day in 1905 Ethel invited them both to tea to meet her family. No doubt to an only child, the Murphy family of four brothers and three sisters must have seemed enormous, especially as they were all encouraged to bring their friends in to Sunday tea. The room seemed to be packed with young people, all talking excitedly, constantly teasing one another, yet full of mutual pride and affection.

The eldest brother, Arthur Murphy, was working at that time in the head office of the Capital and Counties Bank (later Lloyds Bank) in Threadneedle Street, where in fact he remained until his retirement. Leonard, the second brother, was regarded as the brilliant one of the family. An electrical engineer in the Post Office, he was immensely inventive and full of original ideas. After the World War I, he founded his own company to manufacture electric motors, and he pioneered an electrically-driven car, the fore-runner of our present "milk-floats". He also developed a prototype electric vacuum cleaner. In between Arthur and Leonard came Evelyn, a teacher, and after Leonard, Winifred, also trained as a teacher. Then came Harold, whose blond good looks made him a natural choice in "drag" in family dramatics, but with less ability and drive than that of the others. He worked in the Telephone Department of the Post Office. Hilda was particularly attracted by the fourth brother, a serious seventeen-year-old called Frank, then at college. Finally, there was Ethel herself, a lively and intelligent girl whose good looks and sharp wit made her stand out in any company. Unlike her sisters, she did not want to teach. She had a good voice, and was encouraged to take lessons in piano and singing. A concert-pianist friend got her a post as tutor-companion to the daughter of a wealthy Hungarian family, and she spent two most enjoyable years in Hungary, learning German and Hungarian, joining in the elegant social life of the Reviczky family, and meeting a Hungarian artist, Josef Karsay, whom she eventually married in 1913. There had been a special bond between Ethel and Frank during their school days and right up to the time she went to Hungary. This made it hard for her to accept Hilda as Frank's chosen companion and the object of all his confidencers. Nevertheless, Frank retained his affection for his favourite sister, and did his best to help her in later years.

Between 1908 and 1913, all seven Murphy children had married. They set up homes, all within easy reach of old John James and Annie Murphy, in the expectation that life would continue along its usually smooth course. But in 1914 The Great War broke out, and things were never to be the same again. Arthur the eldest Murphy boy and Harold the third son both went into the Army; Leonard with his poor eyesight failed the medical and remained in his civilian job; Winifred's and Evelyn's husbands went into the Navy; Josef Karsay, the Hungarian who had married Ethel, was advised that both he and his wife were now "enemy aliens", so it was thought wise for them to emigrate to the United States. During the first two years of the war Frank remained in the Post Office, as his post was classified as a reserved occupation, but by 1916 he was released for military service and in rather amusing circumstances came to enlist in the Royal Flying Corps, then a new and very modest branch of the Army.

At first, Frank had volunteered for the Signals Section of the Royal Engineers, but the Major who interviewed him (over drinks in the Strand Palace Hotel) decided that as Frank could not ride a horse, he was no good for the Royal Engineers. Frank returned to the Post Office and reported this to his Assistant Engineer-in-Chief, a very humorous and intelligent Irishman, who asked Frank why he hadn't tried for a commission as a wireless officer in the Royal Flying Corps. "Because I don't know anything about wireless", said Frank. To which his chief replied, "Nor does anybody else, so what?" Accordingly Frank went for a formal interview at the Air Ministry headquarters, and was offered a commission as a Second Lieutenant. Asked rather casually at the end of the interview whether he would prefer a General List or a Special Reserve Commission, Frank confessed that he didn't know, and asked the Major for his advice. Fortunately the Major recommended the Special Reserve, which Frank found, when he was eventually demobbed, carried a gratuity of £450, instead of the £110 he would have received, had he opted for General List.

For the first few months Frank was kept busy teaching "wireless" to new recruits. This consisted of getting them to practise receiving Morse code signals from the Eiffel Tower on early valve sets. To get these working, the operator had first to use a match to warm up the pip on the end of the valve.

Towards the end of 1916 Frank was drafted to France as a Squadron Wireless Officer with the British Expeditionary Force. In Frank's artillery observation squadron, his particular job was to look after both the wireless transmitters on the aeroplanes and the receiving sets at the artillery posts. The planes used were Bristol biplanes, at first BE2C's, and later BE2E's, and to spot the enemy artillery posts they had to fly perilously low over the enemy lines. Communication between pilot and navigator was not easy; and on one occasion Frank was almost too late to make his pilot realise that their port wing was on fire. Fortunately, they did get down in time.

Frank did not question the validity of the Allies' war policy, (in public, at any rate). What he did criticise was the exasperating waste and inefficiency with which operations were carried out. "Only twice did I make the mistake of going by train. Once, on my first arrival in France; and secondly, after the Armistice, when the first train which ran right through from Boulogne to Cologne took twenty four hours to travel 250 miles. . . . Another ridiculous side to transport was that in England drivers crept along with headlights so dimmed they were like a couple of glow-worms, whereas in France you went tearing up the roads leading to the front line with headlights full blaze."

Another big problem was leave, or rather, the lack of it. Unlike the pilots, who were usually granted home leave every three months (if they lived that long), the wireless officers were kept in France, as the authorities didn't seem to think they needed any leave. Frank did not envy the pilots, however. As he said, "The average life of a pilot in France was about six weeks, and no wonder, since as often as not they were being sent out with only six hours' flying instruction". Someone back in England, however, must have been convinced that this new form of communication held greater possibilities beyond reporting the range of enemy artillery, for after his eight months in France without leave, Frank was recalled to England and told to set about forming an Officers' Wireless Training School at Farnborough in Hampshire. Delighted at being given a formidable challenge, Frank set about his new task with characteristic energy and thoroughness. The Army had allocated a very old building at Farnborough, known as the Blenheim Barracks, which he thought privately must have dated back to the Crimean War, but otherwise nothing existed.

So Frank decided to run courses of twenty-two weeks instruction, with an examination each fortnight, and wrote the syllabus himself and chose the wireless instructors. His pupils were a mixed bunch, varying in rank from humble second lieutenants to ex-infantry majors, and Frank and his staff crammed them like chickens and hoped that when they had to return to France, some of their new wireless knowledge would remain with them.

Meanwhile, Frank had found a pleasant house called Hartwell Lodge at Farnborough Green for himself, Hilda and their young son Kenneth. Nine months later, a second child was born, on 4th June, 1918, a daughter, who was to be named Joan and given the second name of Ethel, after Frank's favourite sister.

The Wireless Training School was divided into three sections—officers, operators and mechanics—under the administrative charge of an Army Colonel. He and Frank did not get on too well, though he respected the work Frank was doing. His ambivalent attitude was well illustrated when the Armistice was declared in November 1918. On the one hand, he saw that Frank was recommended for the award of an MBE for his work in organising the officers' school, but on the other hand he dashed Frank's hopes of an early demobilisation by getting him posted overseas again to France.

After a few weeks, however, Frank was appointed Commanding Officer (Wireless) for the aerial post operating between Folkestone and Cologne. His job was to equip the planes with two-way radio-telephone sets, and similarly the ground stations between England and Germany. Regular ground-to-air communication along the route enabled the pilots to get up-to-date information on weather conditions. When Frank took over, the range of the radio-telephone sets was only 30 miles, but by redesigning the circuits, he was able to improve the range to 250 miles, so that operators in Folkestone and Cologne could talk directly to each other. Years later, his comment was, "The first use of this improvement was typical – the Army 'brass hats' used it to send loving messages to their wives!"

Though not yet a civilian, Frank spent a most enjoyable twelve months in Germany after the Armistice, despite the Colonel. The Royal Air Force had thoughtfully provided him with an official car, which enabled him to tour the beautiful Rhineland scenery, and as he also had an aeroplane at his disposal, he was able to fly home to England nearly every weekend. (By this time, Hilda and the children had moved to Bromley in Kent, which was rather more convenient than Farnborough for Frank's weekends at home.)

As 1919 came to an end, the time for Frank's demobilisation approached, and with it the fateful question of the choice of a civilian job. He described his dilemma in this way:

"As a married man with two children I needed security,— but my complaint about the Civil Service was that it gave you too much security. Short of murdering your boss, you were unlikely to get the sack. . . . I was offered a radio research job with the RAF. This was very tempting, as I like research work. But I couldn't see myself able to educate the children in the way I wished, on the pay attached to the job. . . . My next idea was that I should get the security I needed if I obtained a post with a big firm. Then I realised two things: one, that even if I did get such a job, my boss would only have to have the wrong thing for breakfast and I should be out on my ear; and the other, that if a man's security was not inside him, it was not there at all. . . . So I decided that with my £450 demobilisation gratuity I would start my own business."

But the question still had to be decided—what business?

Chapter Three

THE PUBLICITY BUSINESS

"A man may write at any time, if he sets himself doggedly to it."—*Samuel Johnson*

The prospect of returning to his old job in the Post Office after the War held out no attraction for Frank Murphy, as one might expect with a man who had experienced the hazards of flying and the satisfaction of successfully setting up a wireless school. There were no comparable challenges in the Post Office, though it did offer secure employment and the ultimate prospect of a pension at a time when many ex-officers were desperately trying to find some way of earning a living.

Mindful of his family, Frank Murphy did go back; but within two weeks he was itching to get out again, with all his worst fears confirmed. He had no idea what else he could do, except that it must give him the opportunity to exercise responsibility and initiative.

One day he happened to be grousing about life in general to Leonard Goss, a friend from R.F.C. days, who was now in the printing business. This reminded Leonard of another discontented fellow he knew—his brother-in-law, Charles Rupert Casson. Casson, who had spent the war as an officer in the 10th Royal Fusiliers, was, like Frank, a survivor from some tough experiences. A bachelor with an engaging wit, a flair for amateur acting and writing, and no other obvious qualifications, he too was determined to do something more positive with his life.

"Why don't you and Frank get together?" suggested Leonard.

Before long, the three of them were deep in discussion of possible projects. It was Frank, the trained engineer, who came up with a mad, but faintly possible idea. He pointed out that many engineering firms were not achieving their market potential because their advertising was hopelessly inadequate. "What about starting a publicity service for them?" he said. "We ought to be able to do it—I can handle the technical side, Rupert can write the copy, and Leonard can do the printing for us." So the "ENGINEERING PUBLICITY SERVICE" was born in January 1920, on the modest capital of £2,000, half of it from Frank's and Rupert's war gratuities, and the rest subscribed by various trusting friends and relatives. The two partners rented a tiny upstairs room in Kingsway, which was the street where most of the engineering firms had offices, and boldly put up a door plate marked "E.P.S."

Rupert Casson had since confessed that he did not tell Frank how little he knew about office routine. Indeed, no sooner had he seen Frank safely out of the door than he hurried round to the local library to take out a manual on book-keeping and accounts, which he had a vague idea might come in useful. Fortunately, Frank's wife, Hilda, with her sound secretarial training, was able to provide some background order to this rather crazy enterprise.

They opened their sales campaign with a mailing letter to all the firms showing at the Motor Boat Exhibition which happened to be on at that time, assuring them of the brilliant prospects awaiting them if only they would let this unknown and untried agency handle their advertising—with no response.

So they continued preparing advertising schemes, and Frank trudged up and down Victoria Street, calling at every engineering office and trying to persuade the firms to offer them their advertising, but although they all expressed interest, no one actually gave them an account. After many fruitless weeks, the capital was rapidly disappearing, and by the end of twelve months they were down to their last few pounds.

Suddenly, Frank thought of another approach. He had noted that, because of the excess profits tax, many firms were spending large sums of money having photographs of their products retouched rather than pay money to the government in tax. He made a calculation that in London at that time, some £10,000 worth of retouching was being given out every month, and that all E.P.S. needed was £400 a month to put them on a profit-making basis. Fortunately, they had built up a good retouching studio, in the charge of one Stanley Knowles, so Frank put on his hat and coat, abandoned all highbrow selling schemes, and went round again asking the firms if they wanted any retouching done.

The answer was a very quick "Yes" or "No". If it was "No", within thirty seconds he was on his way to the next firm. But more often it was "Yes"; and so within a month he had collected the necessary £400 worth of retouching business.

Confidence restored, Frank landed the first account. It was with a company named Elwell Limited, whose business was manufacturing tall wooden masts, an essential back-garden accessory for domestic wireless reception in the 1920's. Frank's knowledge of wireless now gave E.P.S. an edge over their competitors in obtaining the accounts of radio firms, and they signed up several more. They were particularly anxious to handle the advertising of Mullard Valves, and Rupert Casson tells the story of how, having taken the Managing Director out to a dinner they could ill afford, they were elated when he finally agreed to give them a trial. Knowing that it would be his task to write the copy, Rupert asked, "What would you say distinguishes the Mullard valve, and in your view makes it superior to all the other makes?" There was a long silence, and then Mullard said "Well, nothing really". Such honesty was refreshing but hardly helpful. Undaunted, Rupert replied, "Right. We'll launch the campaign on the slogan—"Mullard—the Master Valve!"

In December 1920, E.P.S. landed another client—Ediswan Lamps. This was quite a sizeable account, and Frank and Rupert judged that it called for an advertisement not only in the technical journals, but in the national daily press. It was booked to appear on the day after Boxing Day, so that morning Rupert hurried downstairs to feast his eyes on his first public masterpiece. There it was, but, oh horror!—at the foot of the advertisement was printed in bold type: "THE EDISON ELECTRIC LIGHT COMPANY"—which was not the name of the firm at all! Despite this shocking gaffe, Ediswan did retain their services, and the office-boy who typed the error was not given the sack.

In 1921 E.P.S. managed to secure two more big clients— Belling Limited (electric fires and cookers), and Armstrong-Whitworth Limited, the well-known general engineering firm. Other accounts followed, and in 1922 E.P.S. decided to open their own Art Department, and took on several artists, among them "Willie" Wall, and also Ethel's husband, Joseph Karsay, the Hungarian.

The new department, in a separate studio in Poland Street, did well enough, but disputes developed within the firm, and eventually Karsay moved off with another employee, C. C. Stanley, to form their own agency, ARKS PUBLICITY. Later, Stanley took over Cambridge Scientific Instruments, one of their clients, and formed a company which became famous as PYE RADIO LIMITED.

Meanwhile, the Karsays had to return to the United States, where they had lived during the war, as otherwise Karsay would have lost his American citizenship.

By 1924, E.P.S. were doing well enough to move to Hart Street (now Bloomsbury Street), and since they had enlarged their scope to handle non-technical as well as purely engineering accounts, the name of the firm was altered to MURPHY CASSON LIMITED.

Between 1924 and 1928, MURPHY CASSON grew steadily in size and reputation, and the Murphy family were able to buy a newly-built detached house at the foot of Hendon Avenue, North Finchley. They called the house "Madron", after the Cornish village where Frank's mother, Annie Leggo, had been born. The children were growing up. Kenneth was travelling by train to Highgate School, while Joan attended the local Convent School, and then at the age of eight was admitted to Queen Elizabeth's Grammar School at High Barnet. In 1925 a third child, Frank Maurice, was born, rather to the surprise of his parents, who had thought their family was completed.

It was a happy time for them all, and especially for Hilda, as the years of wartime anxiety and post-war struggle receded. There was much entertaining of friends and family relatives. Kenneth and his cousin Grace (Leonard's eldest daughter), and Joan and her cousin Dennis (Harold's son) had particularly close links. Frank and Hilda joined the local tennis club; the children and their friends played in the Dollis Brook (then still in open country), learnt to ride bicycles, and tried their best to spoil baby brother Maurice. The family annual holiday was spent touring in the Morris "bull-nose" open four-seater.

This typical, almost idyllic suburban existence might well have continued for another ten years, if Frank Murphy had not become dissatisfied with his chosen career. Some things about it he liked very much. For instance, he enjoyed the challenge of seeking out and capturing new clients; and also the never-ending search to find new ways of conveying the "message" of their products. But, by definition, the clients had to have the last word on what was printed, no matter how many attractive ideas Frank conceived and Rupert formulated, and as important clients such as Mullard frequently threw out copy that the two had spent many hours on, Frank became increasingly frustrated. Moreover, he was not using his training as an engineer or his specialist knowledge of wireless, except as a hobby at home.

By the late 1920's, wireless had passed the "cat's whisker" stage of individual enthusiasts, to become, through the medium of loudspeakers, the source of family entertainment. There was a rapidly spreading interest in the "miracle" of wireless as the British Broadcasting Company, later the Corporation, began under Lord Reith's direction to broadcast regular programmes of recitals and talks. However, as Susan Briggs has so truly written in her book *These Radio Times,* the knowledge of how sound was carried by wireless waves and captured by domestic receiving sets was a mystery to the vast majority of the British public. To supply the demand for sets, a large number of companies sprang into existence, some of whose reputations were, to say the least, dubious. If the sets on which they had promised "amazing reproduction" went wrong, the buyer had no redress; the fly-by-night firms were concerned only to get the customer's money, not to service faults—naturally enough, as no one had much idea what caused the faults, or how they could be avoided.

Bigger companies such as Marconiphone did produce better quality and more reliable sets, but they tended to be too expensive for the ordinary working class family.

With the background of his own knowledge, Frank Murphy was convinced that designing radio sets was not a mystery, but a science, and if a science, then subject to measurement. "In fact, if we could only measure what was going on inside a radio set, we should be able to design one on the drawing board, knowing full well that if the instructions were followed, the set would not only work, but work with the expected performance."

Furthermore, he had recently acquired a copy of Henry Ford's book, "My Life and Work", and had been greatly impressed by it. He was particularly struck by the success of Ford's unorthodox policy of paying his workers not the lowest, but the highest wge he could afford. Could such a policy work in England? Frank

Murphy believed that it probably could, but there was no point in trying to prove it in a business like advertising, where the clients dictated the budget. The only course was to attempt it in a primary occupation, such as manufacturing.

So Rupert Casson was not entirely surprised when in 1928 Frank announced that he wanted to leave Murphy Casson and found his own radio manufacturing business—"to give people what they want—value for money".

Frank had already canvassed the opinion of a number of his friends, and found them distinctly discouraging. One of them, who had been a fellow instructor at the R.F.C. Wireless School, was having lunch with Frank one day in Decenber 1927, and was startled when suddenly asked, "What would you say if I told you I was going to chuck this business and go in for making wireless sets?"

"I should say you were a bloody fool!"

"Nevertheless, though I value your opinion, I still propose to go ahead . . ."

"In that case, I should still say you were a bloody fool!"

Six years later, when Murphy Radio was triumphantly established, he had the grace to admit he had been wrong.

Another man whose opinion Frank Murphy valued was the former Assistant Engineer-in-Chief of the Post Office—the same one who had originally advised him to enlist in the R.F.C. Asked what he thought, he shrugged, and said, "You're seven years too late". It certainly looked as if he was right, as many big firms were already well established in the market.

Above all, Frank had to convince his wife Hilda. He was asking her to sacrifice their present comparative affluence for several more years of struggling to make ends meet, and now they had three children to provide for and educate, with no guarantee that they would not lose everything. Frank's current salary as a director of Murphy Casson was in the region of £2,000 per annum, and he did not intend to draw more than £750 per annum until the new business was secure. It was a hard and difficult decision to make.

Eventually, after many hours of discussion, Hilda's faith overcame her doubts. She, more than anyone, knew what Frank could achieve once he had set his mind to it. The formation of the Officers' Wireless School and the building up from nothing of the advertising agency bore witneess to his technical and administrative ability. Besides, once Frank had decided that a certain course was right, he was determined to pursue it at all costs, and to grind away all opposition. Thwarting him, even from the most sensible of motives, would only lead to friction and unhappiness. So Hilda quietly told him to go ahead.

There remained the technical problem of how to withdraw Frank's share of the advertising business (calculated by mutual agreement at £4,000), in order to provide working capital for the new venture. Payment of a lump sum would have caused serious problems for Murphy Casson Limited, but as the initial needs of the embryo radio company ere not likely to be large, Frank was content to withdraw the capital in small instalments as he needed it, a mutually satisfactory arrangement.

Chapter Four

THE BEGINNING OF MURPHY RADIO

"Nothing great was ever achieved without enthusiasm."
Emerson.

One day in 1936, when his business was acknowledged to be one of the leading companies in the radio industry, Frank Murphy was asked to give a talk at a public gramophone recital organised in Dover Town Hall by the local Murphy dealer. He chose as his subject: "How I came to start Murphy Radio". The following quotation illustrates both his moral and his scientific approach.

"Before committing myself, I had to decide whether the making of radio sets was in itself a worthwhile occupation, as I felt that it was only with a commodity which was valuable from the community's point of view that I was likely to succeed in permanently paying higher wages.

The first conclusion I came to was that entertainment had a definite niche in human life and that, although there are those who argue otherwise, human beings are just as entitled to have entertainment as to have work. Further, radio entertainment particularly appealed to me, as a means of placing first-class entertainment within the reach of the great majority of people. My argument was, people with "Class A" incomes are able to take care of themselves, as their resources automatically put every variety of entertainment within their reach. On the other hand, even today, people in the Class B and C groups—the former because 'keeping up appearances' absorbs so much of their income, and the latter because their means are so limited—have only forms of cheap entertainment open to them, namely, reading (via the libraries); dance-halls; cinema and radio.

The case seemed to be therefore pretty clear, and although commercially I was an unborn babe when I started Murphy

Radio, I felt it was an asset rather than otherwise."

It was an unusual approach, to say the least. While, like Henry Ford, Frank Murphy did not deride the profit motive in business, his primary aim was to bring enjoyment within the reach of the vast majority of ordinary folk. Or, as he would have said later, to raise their standard of living.

Back in 1929, however, such a long-term ideal was likely to be submerged in the practicalities of daily living and backroom research. Frank had persuaded a family friend, Edward J. (Ted) Power, to join him in his enterprise. He, like Frank, already had several years' experience of wireless theory and practice, having been a radio operator in submarines during the war, after which he had joined the McMichael Radio Company.

The problem was, not how to build a radio set—that was within the range of many competent radio enthusiasts—but how to make one which was a sound engineering job, simple to manufacture and to operate, and above all, reliable. It is difficult to realise that, up to 1929, a "wireless set" usually consisted of a number of separate items, i.e. loud speaker, headphones, receiving apparatus, battery or accumulator. Even when enclosed in boxes, the apparatus was cumbersome and vulnerable—especially if the conscientious housewife flicked a duster over her sideboard where "the wireless" normally sat. As an object, it could hardly be called a thing of beauty, and to most families it was regarded as "Father's toy", and more likely to infuiriate than to entertain them because of its unpredictability. A humorous article by Beverley Nichols, which appeared in the *Radio Times* as late as June 1931, illustrates the way in which this mysterious phenomenon was regarded:—

"Father, kneeling before a wireless machine enclosed in a cabinet of particularly revolting Jacobean design, is fiddling with it. Sucking a large sweet, his small son gravely regards him.

Son: What are you doing, Father?
Father: Trying to get Moscow.
Son: Why?
Father: Never mind why.
Son: But I thought you said Russia ought to be kept out of England?
Father: (excited) What was that noise? (A shrill shriek

comes from the instrument).
Son: Oooh, Father! Was that Moscow?
Father: Atmospherics.
Son: Is Atmospherics a Russian station?
Father: (pompously). You would not understand what atmospherics are.
Son: What are they?
Father: A . . . . . . .a disturbance in the ether.
Son: What's a disturbance in the ether?
Father: Er . . . . er . . . . waves. Er . . . . waves . . . . (Another shrill screech comes through, accompanied by some unintelligible words).
Son: Ooh! The Bolsheviks are having a battle!
Father: Nothing of the sort. It was Daventry having a Welsh interlude."

Frank Murphy and Ted Power were determioned that their first set would be very different. First, it would be guaranteed to work; second, it would be realy simple to operate; third, it would have excellent reproduction; fourth, it would be a pleasant piece of furniture in its own right; and finally, the price would be such that it would be within the reach of the vast mass of ordinary families, who would be genuinely getting "value for money". To achieve all this, and to lay the foundations for a solidly based company capable of fulfilling its sales potential, took them eighteen months of unremitting effort.

The first thing was to decide the location of the factory where the sets would be made. Rather than North London, Slough, or the Great West Road industrial estates, Frank Murphy settled on Welwyn Garden City, a new development which offered not only well-serviced industrial sites, but also excellent housing for factory employees, in an attractive woodland area of Hertfordshire. The Garden City was then only nine years old, and was popularly reputed to be inhabited mainly by cranks wearing sandals and flowing robes. The Welwyn Garden City Company, set up to carry out Ebenezer Howard's ideas and managed by C. B. Purdom, Richard Reiss and Frederic (later Sir Frederic) Osborn, had attracted the famous firm of Shredded Wheat, which had built a striking modern factory and silo alongside the railway; and there were one or two engineering firms such as

Dawnay's and the Norton Grinding Wheel Company. Smaller firms were being encouraged by a number of small sectional factories tucked away in side roads. One of these, 10,000 square feet in size, was rented by the newly-formed "Murphy Radio Ltd."—and, much to the family's amusement, the adjoining factory was let to "Bickiepegs", a firm making rusks for babies.

By then the two originals had been joined by S. (Sidney) Carne, later to become the first Works Manager, but at this time merely Laboratory Technician and Tester. He also had to see to the more mundane tasks of cooking, washing-up and buying provisions for their mid-day meal. Four years later, in the first edition of the *Murphy News*, Carne described his working day and that of the others, in light-hearted terms.

"My jobs embraced the manufacture of new instruments, measuring band widths or inter-electrode capacity of valves and other 'lab' work; while the others filled sheets and sheets of paper with Chinese and other hieroglyphics and each grabbed for the slide-rule as soon as the other put it down. At about 11 a.m. it was necessary for me to do the shopping. This was made easier by having only one shop to to (Welwyn Garden City is like that!).

Another short intermission at work, then the business of housewife started in earnest, and in due course the gallant efforts of the cook, often accompanied by pointed remarks from the others, produced a meal of meat, vegetables and a sweet.

Incidentally, the sweet was supplied in turn by our three wives, and when driving up in the morning it was somebody's stern duty to see that the pudding was neither spilt during the journey nor kicked out when we arrived.

Anyhow we never had any food left, doubtless due to one of three causes: (a) it was cooked so well, (b) people will eat anything when they are hungry, (c) there wasn't enough.

After washing the pots and stoking the boiler I went back to the band-widths and similar toys . . . And so to teatime, with rolls and butter and fruit, and after that it was tidying up, stoking up the boiler for the night, and then back down the Great North Road to warmer climes . . ."

Some indication of the scope of the experimenting and testing in these early months can be gleaned from a rather more sober source—a pamphlet which Frank Murphy issued in 1930, modestly entitled: *Some Notes on the Design of a Battery-Operated Portable.* Headed characteristically, "Making Wireless Simple", the article begins bluntly:

"The design of a battery-operated portable must be based on certain fundamental considerations, if satisfaction to the ultimate customer is to be assured. These are:—
 (1) Amplification
 (2) Selectivity
 (3) Ease of control
 (4) Uniformity of performance
 (5) Reliability
 (6) Quality of reproduction
 (7) Appearance
 (8) Price
 (9) Economy of operation
 (10) Weight

As in every engineering problem with numerous considerations to be met, the answer finally adopted is in some measure a compromise. This should be clearly recognised, as through years of propaganda the wireless buyer has been led to expect 100% performance in every direction. This is of course to be striven for, but it can never be attained."

The writer then goes on to discuss each of the ten aspects he has listed, giving a simple and refreshingly candid explanation of how and why the exact specification of the Murphy B4 Portable was arrived at. Most of the pamphlet is of interest to the technical rather than to the general reader, but there is a revealing paragraph under the heading "Ease of Control".

"This is of considerable importance from the public's point of view, and it is the writer's personal opinion that today the majority of potential customers for radio neither know anything about it, nor do they wish to. They just want 'the wireless'. . . . As we are blest with only two hands, it is evident that the number of controls requiring simultaneous operation should not exceed two. . . ."

In fact, the new Murphy wireless set needed only one tuning knob to find the various stations. As a further innovation, the dial-scale was marked, not in degrees, but in actual wavelengths. Since the daily papers printed the wavelengths of stations at the head of their daily programmes, it was far simpler for the Murphy owner to find his favourite station than for those who owned sets still marked in degrees.

Valves in radio sets have long since disappeared, but in 1929 good, reliable valves were essential. Frank Murphy gave a detailed analysis in his little booklet of the performance of various types of valves, and it is clear that he and Ted Power had done a great deal of work on this subject, including some original research on the grid anode capacity of screen grid valves.

The page on "Quality of Reproduction" begins with a typical Frank Murphy remark, foreshadowing many a subsequent Murphy Radio advertisement:

"Clear reproduction is important, as, when all is said and done, a wireless set is bought to be listened to. Many claims are made for perfect reproduction, but no such thing as perfection is possible; all that is possible is to approach nearer and nearer to perfection."

Similarly, on the subject of "Reliability", there was no attempt to dodge the problems:

"Reliability is of first importance to public, trader and manufacturer alike. To the first, lack of it means annoyance and general discontent; to the second, expensive service visits and dissatisfied customers; and to the last, loss of prestige, and eventual failure of business if the lack of reliability is sufficiently pronounced. The scale of achieving reliability would seem to be the percentage of sets sent back for service, and in the case of the Murphy, this is five per cent—and in the majority of such cases, the fault is a microphonic detector valve, over which the set-maker's only control is that of rejection on test. Unfortunately, valves appear to become microphonic in transit, and with the course of time. Valve-makers are, however, fully aware of the problem, and there is no doubt that they are always studying valve construction with a view to producing a really non-microphonic detector."

The booklet concludes with a half-page illustration of the Murphy B4 Portable, and with some photographs of the Murphy Radio factory in Hyde Way, Welwyn Garden City, showing the press shop, assembly lines and test benches. (See illustration, Fig 8.).

There is another first-hand account of the intensive testing to which the prototype B4 was submitted. It was given by Sidney Carne, who described how the three pioneers set off in Frank's car one morning for Cornwall. Their intention was to measure the field-strength of the existing transmitters at varying distances from them, also to test the selectivity of the set in obtaining station signals without interference under very severe conditions, e.g. close proximity to another transmitter. It says much for the reliability of the car (a Ford, needless to say) and the endurance of the driver and his passengers that they reached Bodmin—the furthest point for testing—shortly after lunch. Returning, they continued to stop at intervals for measurements, calling at Shaftesbury for dinner, and home in London in the early hours of the following morning, having driven 500 miles in 20 hours.

To keep their finances afloat during the long months of research, the three pioneers offered to make some sets for Gamages, the famous London store. These sets were marketed as the "Gamages Popular Two" and "Gamages Popular Three", and were good of their kind, but not outstanding for technical innovation. Producing these sets, however, provided valuable experience in organising a small work force in a machine shop and along assembly lines—most useful when the day arrived for 'tooling up' for the first real Murphy set.

Prior to this, and about six months after the daily trek from Finchley had begun, someone else had made a decision. Hilda Murphy increasingly felt that the strain of driving forty miles each day on top of long hours of research was too much to ask of her husband, and she decided that the family must move to be nearer the factory. Within a week she had found a house to rent in Welwyn Garden City, at number 25, Brockswood Lane (see illustration, Fig 11.) and in the autumn of 1929 the Murphy family said goodbye to their home in Finchley, with the inevitable loss of friends and change of schools.

One advantage of settling in a new town is that most other people are new, too. It was impossible not to make friends when everyone had to do their shopping in the one and only shop—which, for quite a number of years, consisted of a large shed reminiscent of an aeroplane hangar, and just as draughty. Welwyn Garden City in 1929 had only one public house—the "Cherry Tree"—which also served as restaurant and dance hall.

(At least, it was better than Letchworth, the earlier Garden City, which did not have a pub at all.) What Welwyn did boast was a really large cinema, which doubled as a theatre, and there was also a tiny "barn" theatre. Consequently there were at least half a dozen amateur dramatic societies whose performances were professionally polished. Dame Flora Robson appeared on the Barn Theatre stage, as a young amateur actress. C. B. Purdom wrote and produced plays there. The influence of George Bernard Shaw, living at nearby Ayot St. Lawrence, was very evident. And not only drama, but many other clubs and organisations flourished in the Garden City, particularly Scouts and Guides.

It was not long before Hilda found herself drawn in to do reception work at the local Baby Clinic and the Red Cross branch, and very soon she was persuaded to join the Girl Guide Movement and to take on the post of District Commissioner. She took this job seriously and did it very well. She had a knack of getting on well with children of all ages, but she had a special fondness for Brownie enrolments, when she was always called on to "tell us a story!" She found immense fun and satisfaction in her Guiding, and somehow managed to squeeze it in with her other activities of running a household, bringing up the children and, above all, supporting her husband. This last included listening to, and often challenging his ideas at home, and giving practical typing and secretarial help in the tiny works office. As a Director of Murphy Radio, Hilda believed in pulling her weight.

Chapter Five

"NOW WE'LL TELL THEM!"—
THE MURPHY ADVERTISING.

"I will a round, unvarnish'd tale deliver . . ."
                                    Shakespeare: *Othello*.

Frank Murphy was sure that the general public would want to buy Murphy sets, once they had heard them; but the problem was, how to tell people about them. To the mind of a former advertising agent, there was only one answer.

In the early days after leaving Murphy Casson Ltd., Frank would occasionally call on Rupert Casson in his London office to see how the old business was going—(none too well at that time, as one of their major clients had gone bankrupt, inconveniently leaving Murphy Casson with a considerable sum in bad debts.) On one of these occasions Frank had said loftily, "By the way, I shan't be coming to *you* to do my advertising; you're not big enough."— which was a remark that Rupert never forgot, because at the time Murphy Radio consisted of only three people, who had nothing ready for the market at all.

However, after vainly approaching a number of the bigger advertising firms who remained politely indifferent to his needs, Frank reconsidered his decision and went back to Rupert Casson, who was then faced with some difficult problems. As he said, "Why should the public be at all interested in this new set, which *looks* much the same as many others which are much better known? *I* think the most interesting thing about Murphy Radio is *you*, Frank Murphy, and your theories of making wireless simple and reliable!"

Following this line of thought, Rupert came to two more conclusions. The first was, that from the start Murphy Radio must appear *important*—which ruled out tiny advertisements (all they could afford) in national daily newspapers and trade papers. Instead, he suggested they should use the whole advertising budget on full page advertisements in the *Radio Times* every fortnight. (At that time, the *Radio Times* was still comparatively cheap.)

Secondly, since Frank's personality and views were to be the keystone of the advertising, a large photograph of Frank with his pipe would occupy virtually all the space, thus reinforcing the idea of solid "no-nonsense" reliability embodied by the Murphy set. This idea was not an advertising gimmick—Frank had always smoked a pipe since his Flying Corps days, and photographing him without it would have resulted in a very un-natural likeness. The written copy would be a mere three or four lines under the photograph, saying that Murphy sets were simple to use and designed to give years of trouble-free service. Several trial proofs of this kind were knocked up and delivered to Frank, who gave them one look and said, "Nauseating!" He told Rupert sharply that he had no intention of being made the focus of the advertising.

So the first advertisement appeared on October 3rd 1930 in the *Radio Times* without Frank's picture in it (see opposite page 36). Instead, there was an artist's sketch of a hand operating the tuning dial marked in wavelengths, with the main heading "MAKING WIRELESS SIMPLE." After this came quite a lengthy piece of copy, explaining how simple the set was to operate, and giving the price and a technical description of the set. At the bottom of the page was a coupon for a free demonstration (which the reader was invited to mail to Murphy Radio), and a photograph of the new Murphy battery portable.

There were four more advertisements in the *Radio Times* during October and November 1930, in one of which Frank appeared, although the copy was written in the form of a personal message and signed with a facsimile signature. In all of them the principal theme was "Making Wireless Simple", illustrated by a pretty little four-year-old girl triumphantly finding a station on the set. The returnable coupons continued to appear, but also, significantly, a list of the names and addresses of "appointed Murphy Radio dealers", who could demonstrate and supply the sets.

The arguments with Rupert Casson went on through the winter, while Frank and his first representative, Arthur Herod, were driving into town after town sizing up the radio dealers and

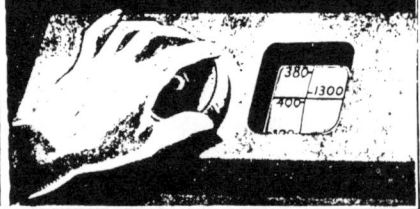

MAKING WIRELESS SIMPLE

Look up the wavelength of the station you want
Set the tuning dial to the same figure
—there's the station !!

THAT'S simplicity isn't it? No fiddling with knobs and trying to remember how you got the station last time. We say "mark the dial in wavelengths because it makes it simple to get any station." We say "only one control for tuning because it makes tuning simple." A Murphy set is a simple set, but it is also a great deal more. It is a beautiful set; beautiful to see and beautiful to hear. It is a troublefree set because we have so designed it. A Murphy set is meant to give you, and will give you, long years of happiness and pleasure. And we sell it at as low a price as we possibly can—17 guineas – inclusive of everything, of course. I should like you to hear one of our sets. Will you send me the coupon?

Frank Murphy

B.Sc., A.M.I.E.E., A.I. Rad. E., Chartered Elect. Engineer

MURPHY RADIO

THE MURPHY PORTABLE

4 Valve Screened-Grid designed circuits giving simple truly centred Calibrated direct in wavelength. Tuning controls, one tuning control, one reaction control, one wavelength switch, and one combined volume control and on-and-off switch. Selectivity is particularly sharp, volume ample and quality of reproduction particularly pleasing because of its all-round high standard.
Cabinet, embodying concealed hand grips, is of selected fine polished walnut.
Battery rack is of special and neat design.
Chassis is readily withdrawn without disturbing the cabinet.
A turntable is incorporated.
Weight complete is approximately 32 lbs.
H.T. and L.T. batteries and valves are included in the all-in price of

17 GNS

To MURPHY RADIO, LTD.,
Welwyn Garden City,   Herts.

Please arrange a demonstration of the Murphy Portable at my home, without obligation.

Name ....
Address ....

selecting the best of them to be "appointed Murphy dealers". The sets *were* beginning to sell, but rather slowly, because not enough people were yet alerted to their good points. Advertising was suspended, partly to save money, but mostly because no one in the radio trade expected to sell many sets as winter gave way to spring and people's thoughts turned to gardening and other outdoor pursuits.

It was clear that a new advertising campaign was going to be needed in April. Rupert renewed the idea of using Frank as the main focus, and after hours of argument, Frank finally gave in. "All right. It's your responsibility!" was his characteristic last word. One of Frank's unusual traits, this. He would argue long and forcefully against a case put to him by a colleague, but once he had accepted the case as being based on expert knowledge, even though he himself disagreed with it, he would place his trust unreservedly in his colleague, and from then on would defend the scheme against all critics.

So, on April 24th 1931, the first "Man with the Pipe" advertisement appeared, with a small inset photograph of "little Joan", (who was actually the daughter of the photographer) with the set (see page 38). The copy had been reduced by Rupert Casson to the minimum:—

"My sets are simple to use. You switch on—then turn the dial to the wavelength you want. You don't have to be an expert to use Murphy sets. *Ask your dealer for a demonstration today.*"

FRANK MURPHY
B.Sc., A.M.I.E.E. A.I.Rad.E.,
Chartered Elect. Engineer.

The name of the firm appeared in large bold letters across the bottom of the page.

Two weeks later, the next advertisement appeared. Again, a very limited amount of copy appeared.

The response to these advertisements was enormous. To begin with, no agency before this had used the direct appeal of a real person speaking in simple colloquial terms to his readers as equals. Most advertisers relied on artists' sketches and photographs of their *products,* and made large-sounding claims for the excellence of their performance in traditional advertisers' jargon. A typical Marconiphone advertisement of October 1930 reads:—

MAKING WIRELESS SIMPLE

"That's fine—now get me another station Joan....."

Murphy Type B4 Portable
4 Valve Screened Grid. Single Tuning Control—marked in wavelengths. No aerial or earth required. Beautiful walnut cabinet. Self-contained loudspeaker and batteries.

CASH PRICE **17** GNS.
Hire purchase terms to suit all pockets.

MURPHY RADIO, LTD.,
Welwyn Garden City, Herts.
Telephone: Welwyn Garden 331.

•

" . . Slaithwaite can be cut out within a few degrees . ."
— *Dealer in Pontefract, Ref. No. 100*

*T*HERE'S one station, here's the other—and there's no interference. That is because we've made it our business to see that Murphy sets are extremely selective. You never have other programmes interfering with the one you are listening to on your Murphy.
Ask your dealer for demonstration to-day.

Frank Murphy
B.Sc., A.M.I.E.E., A.I.Rad.E.,
Chartered Elect. Engineer.

MURPHY RADIO

"Pioneers in wireless thirty years ago, the Marconiphone engineers are the acknowledged leaders today. Their knowledge and experience are behind every Marconiphone product, giving unfailing reliability and first rate performance. Handsomely designed in the latest styles, these loud speakers are very easy to operate and their tone is faultless . . . ."

In the corresponding Ferranti advertisement, the copy-writer almost runs out of superlatives:—

"The Ferranti Console represents the supreme achievement of radio design and craftmanship. "Superb" is the only word to describe the reproduction of the amazing Ferranti Moving-Coil, Magno-Dynamic Speaker—no radio gives such magnificently brilliant and true-to-life rendering as does the Ferranti . . . ."

Here the aim is to dazzle the reader with the superior expertise of the Ferranti and Marconiphone engineers, while the Murphy advertisements do the opposite, emphasising in simple words and "natural" pictures the simplicity of the sets, and the personal commitment of the managing director. Rupert Casson had the enormous advantage of knowing Frank Murphy personally, so that from years of familiarity he could pick out Frank's very phrases, and their very homeliness not only drove home the immediate message, but conveyed the image of a practical honest enthusiast who wanted to share his knowledge with others.

Within a year the Murphy advertisements had become instantly recognisable by their distinctive style and had made Frank Murphy's face and personality widely known. By 1933, he was a national figure and was referred to (long before Stanley Baldwin) as The Man With The Pipe.

Today it is commonplace for products to be advertised by association with a well-known personality from sport or television, but somehow even the most skilfully produced advertisements lack the extraordinary appeal of those early Murphy pieces.

Rupert Casson was right. It was not only the Murphy sets, but Frank Murphy's ideas which captured the public imagination; and as Roosevelt in America revolutionised the political speech by transforming it into a fireside chat, so Frank Murphy turned the ordinary newspaper advertisement into a vehicle for communicating some highly original and challenging ideas.

Rupert Casson said recently:—

"When whole pages appeared showing this unknown chap smoking, with only a few words about his brand new radio sets, quite naturally they did attract attention—which was my first aim, of course. But *not* favourable attention. I can't recall a single complimentary reaction. Plenty of rude, even caustic ones. It was a writer in "Advertisers' Weekly" who said, "It's possible such advertising will sell pipes. Why it should hope to sell radio sets remains a mystery" . . . Frank told me that he suffered a good deal of leg-pulling himself. But he did not interfere. "It's *your* responsibility."

Casson continued:—

"In a few years' time, when Murphy Radio had become a household word, showing photographs of managing directors in advertisements became almost a craze. Barratt's Shoes was about the only such campaign that lasted. (Mainly because the other managing directors didn't have much to say that was interesting apart from the usual claims that their goods were very superior.) My own theory was—and still is—that advertisers are no different from the people we meet daily. The ones that boast loudest are the ones least trusted."

By 1934, the Murphy advertisements were acknowledge to be among the most successful of their kind. Two statements giving C. R. Casson's point of view were reprinted in the *Murphy News* (July 14th 1934). From these it is clear that the clients he most approved of were those who had complete trust in him and his professional skill:—

"If we were rich enough we would take no business except where we were given 'control' of the advertising.

Our clients must tell us what they can of their products and their markets. In practice, we expect to find out a good deal for ourselves. When we have that 'specification', our clients must also give us an idea of their financial position, so that our plans can be laid within practical limits.

But we don't want clients to tell us how much they are 'prepared' to spend. *We'll* tell them how much the job will cost. This procedure is accepted as natural on the part of the Works. It is quite generally looked on as lunacy and impertinence from the advertising agent.

We don't want clients to tell us how to allocate the money either. That's our job and we know more about it than they do, and though we welcome their help and the special knowledge that they may have, we rightly treat it with caution, for very often the 'special knowledge' turns out to be 'special prejudice'.

We don't want clients to tell us what to say or how to say it. Anybody, from the Chairman to the Office Boy, can criticise our layouts and our copy—and goodness knows they all do! We are glad to hear them (sometimes!) but *we* will decide whether to make any alterations.

In short, we want our client to stick to his job—manufacturing and distributing good products. And having given us the job of making his products known, we want him to LET US DO IT.

Caution. Those national advertisers who, after reading the above, feel compelled to entrust us at once with huge sums of money, are warned that applications for our services will be dealt with in STRICT ROTATION. (*With a slight bias towards those with the most money.)*

C. R. CASSON LTD.,
Incorporated Practitioners in Advertising"

Murphy News readers were assured that Murphy Radio *were* one of the advertisers who had given their complete trust to the advertising agents, "It seems to us that the good results which have followed must be obvious to everybody."

In the following issue (July 28th 1934) C. R. Casson Ltd., were given a further opportunity to expand on their views of the function of advertising:—

"THE TEST OF ADVERTISING IS WHAT IT *TELLS* NOT WHAT IT *SELLS!* Many a clever advertising scheme sells large quantities that would be better unsold. That is not good advertising, but bad advertising; bad for the community and worse still for the profession of advertising itself."

This was a pretty courageous thing for an advertising agent to say. Casson also gives humorous examples of "natural" and "typical advertiser's" styles:—

"A manufacturer speaks to a friend about the new car he has

just brought out:—

'It is good, isn't it? Notice that last hill? Forty, in top—and took it like a bird. Yes, I'm pleased with that engine, and the body looks good, too, doesn't it? Mind you, I think we have all got some way to go yet in streamlining, but this bus is well on the way there'.

But often when he talks to the public, he loses that naturalness:—

'Here is a car of outstanding performance, in which you have speed, comfort and dignity, at a modest price. In this new model we have taken advantage of every scientific advancement in motor car design, and it is difficult to believe that we have not at last reached perfection . . .' "

A much more detailed exposition of the Murphy Radio advertising was contributed by C. R. Casson in the *Murphy News* of November 3rd 1934. Starting by pointing out that Murphy Radio were the largest and most consistent advertisers in the radio industry, he went on to say that, despite this, the greatest number of separate insertions in any single paper in 1934 would still be only twenty-seven. "Now to speak to somebody for two minutes twenty-seven times in a year is hardly overwhelming, is it?" Moreover, the number of people who were likely to have read the advertisements could hardly exceed thirty million, and about a quarter of those were children. Bearing in mind these restrictions, Casson went on to list some of the subjects which his agency had dealt with so far in the national press. These included Reliability, Reproduction, Distribution, Buyer's Responsibilities, Hire Purchase, Cabinet Design, Unemployment, and New Sets.

Yet Casson complained he was constantly being urged to deal with still more topics in the advertisements. To be asked: "Murphy? Who's Murphy? Never heard of him!" was a salutary reminder that a large proportion of the British public was unaffected by Murphy advertising.

"That we do actually secure a good deal of attention *comparatively* is due to the fact that the advertising deals with realities, and not with the usual fantasies and extravagances. We talk (or try to talk) interestingly, of course, but as frankly and naturally as though we were talking privately to individual. So we keep a sense of balance. We *discuss*

Fig 1. Frank Murphy as a young man (1908)   Fig 2. Frank Murphy in R.F.C. uniform (1917)

Fig 3. The Murphy family and friends (1920) (F.M. at centre back)

4. Hilda Murphy with Kenneth and Joan (1923)

5. Frank, Hilda, Maurice and Joan on top of snowdon (1935)

6. Teatime music in camp with a B4 portable (1929)

7. Relaxing after a Girl Guide camp (Hilda and Dorothy)

Fig 8. Test bench (B4's in foreground)

Fig 9. The Welwyn Stores in the early '30's

Fig 10. The Cherry Tree Restaurant Welwyn Garden City

Fig 11. The house at 25 Brockswood Lane

Fig 12. Murphy Radio's new factory (1931)

Fig 13. Ludwick Corner in 1930

Fig 14. Ludwick Corner in 1981

Fig 15. Murphy Radio's first assembly line

Fig 16. Press Shop: Murphy Radio

Fig 17. The A26 Radiogram (1934)

Fig 18. "Dignified and Beautiful" the A240

THE FIRST "MURPHY" FOR TELEVISION

Its very distinctive cabinet is carried out mainly in Bombay rosewood, with a lid of impregnated black pearwood and white sycamore handle. The leather-covered frame for the cathode ray tube rests on a quilted mahogany top, while the knob recesses are also leather lined.

Fig 20. The first Murphy television set (1937)

THE RANGE COMPLETE The Type 28 R...lio-gramophone, in a Walnut and Canadian Cherry cabinet, employs a side-by-side arrangement of the gramophone and radio portions, ... facilitating operation

Fig 19. The A28 Radiogram (1935)

THIS...

... was the birthplace of Murphy television

Fig 21. This was the birthplace of Murphy television

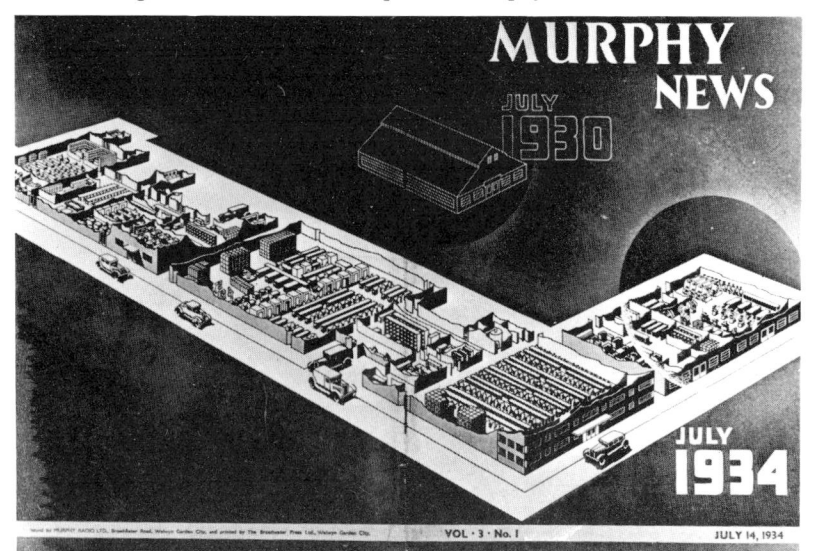

Fig 22. Progress from 1930 to 1936

problems and admit other points of view. We are not thinking always of how soon we can "come to the point" and sell something. We are not always on our dignity, and don't mind printing a joke at our own expense. Murphy advertising is often criticised for these very features. It is called 'undignified', 'indiscreet', 'wordy', 'pointless', 'lacking in sales force'. But I believe that in these things lies its strength. To tackle the subject of unemployment and talk honestly about it; to tackle cabinet design and discuss it boldly and frankly; to talk plainly about cut prices, of the *difficulty* of finding a Murphy dealer, and so on; to drop one's dignity and make a joke about a prize for potatoes—these are natural things for intelligent men to do in private talk.

To do them publicly—in advertising—is both unusual and certain to produce strong criticism and grave doubts. It needs courage to listen to such criticism and still follow your road. And I shall finish by saying that the courage which keeps us all on an unusual and often difficult road, springs originally from our friend 'F.M.' "

(from an article in *Murphy News*, Nov. 3rd 1934)

One of those members of the public who was struck by the advertisements and who took the trouble to write to Frank Murphy about it was a Mr. William Barnes, of Newton, St. Helens. Lancs:—

"Dear Mr. Murphy, I am writing this letter from the outsider's point of view (by "outsider" I mean I am not lucky enough to be the possessor of a Murphy receiver) with regard to the following three items:—

1. The frankness of your advertisements.
2. The quality of your receivers.
3. The courtesy displayed by your dealers.

1. I must confess I have never seen advertisements that speak to you straight from the shoulder and tell you what to expect and what *not* to expect from the goods they are advertising, and therefore eliminating from the would be consumer's mind all shadow of doubt regarding the merits of the goods he is thinking of purchasing. I always look for your new advertisements to see if they all display the same frankness, and have never found them to vary one bit.

WHAT A LIFE !!!

I received these letters within a week or two of each other. Perhaps you would like to read them. —F.M.

Dear Sir,   16 9 33.
I note that in a recent Mail,
You have the awful check to fail
To wear that everlasting pipe;
And so, dear sir, the time is ripe

For me to say a thing or two.
The first, what would Jim Thomas do
If printed in the News or Star
Without that shirt and fat cigar?

Likewise James Maxton's heated flare
Goes down to history per that hair,
Which o'er his forehead loves to stray
And there will be when he is grey.

Or would dear Ramsay's picture " wash "
Without that jaw and that moustache?
For once a style goes into press,
It stays, or there's a blooming mess.

You might as well, dear sir, forgo
The features of your radio,
That's marked for all to hear and see
With an outstanding pedigree.

For when tired man his pipe has lit
And, slippered, in his chair doth sit,
Another pipe comes to his scan—
He thinks of radio's superman.

And having thought, the following day
He's late. For on his homeward way
He calls in at the wireless store
And views all kinds of sets galore.

And to the dealer says—" My hat!
Poor stuff, this make—I don't like that...
Ah! here we are—Yes, that's the type;
Made by the chap who grows a pipe! "

And so, I pray you, stop this " tripe "
And reinstate that national PIPE.
Then you'll be really good, you know—
As good as MURPHY RADIO.—F.W.A. (A MURPHY USER)

Mr. Frank Murphy.   25 10 33.
Dear Sir,
     This tobacco advt. is very bad for the wireless trade generally as well as for your business.
     To unblushingly constantly exhibit oneself as a weak slave to the tobacco habit or vice is no recommendation to your wireless goods, or to your factory where your employees will doubtlessly be imitating you.
     Again, it is an encouragement to the public to continue squandering millions yearly on tobacco with no real benefit to the nation in the end. The tobacco and wireless trades should be kept distinct, as they have nothing in common.
     It encourages the public to spend money on tobacco, whereas the public should be encouraged to spend less on tobacco and more on wireless. This would greatly be to the advantage of the nation.
     Why do you not advertise wireless goods instead of tobacco? Many competent but *very honest* advt. writers could do very much better for you by producing something as good as " His Master's Voice " advt. (fox terrier listening to gramophone.
                                                          Yours truly,
                                                         A FRIEND.

● YOU CAN BUY MURPHY SETS ONLY FROM APPOINTED MURPHY DEALERS

A4 *For A.C. Electricity.*   **D4** *For D.C. Electricity.*
ALL-ELECTRIC 4-VALVE SUPERHET
Single Tuning Control. Moving Coil Speaker. Illuminated wave-length dial. Walnut Cabinet inlaid with Rosewood.
Cash Price **£14·10·0**
Any model on hire purchase.

B5 *Class B Amplification.*
BATTERY 5-VALVE SUPERHET
Single Tuning Control. Moving Coil Speaker. Illuminated wave-length dial. Walnut Cabinet inlaid with Rosewood.
Cash Price **£14·10·0**

MURPHY

MURPHY RADIO LIMITED, WELWYN GARDEN CITY, HERTS.   TELEPHONE WELWYN GARDEN

Being a draughtsman and a designer by profession, naturally the first thing I look for in a machine is precision and symmetry, as well as quality, and I may say that it gives me a real pleasure to run my eye over your receivers, noticing how each detail has been arranged to form a perfect combination.

3. In one of your announcements a while ago you said that your dealers were picked for their ability and courtesy. I have spoken many times to your St. Helens dealer, Mr. Rothery, of Rothery Radio Ltd., and have always received from him that attention and courtesy, even on the slightest detail, that one would expect from a Murphy dealer, and he always delights in bringing to one's notice the good points of Murphy receivers.

In conclusion, I wish to congratulate you on such a fine range of receivers and wish you every success. It must be a pleasure to work in your factory designing and constructing these well-nigh perfect receivers.

WILLIAM BARNES."

Two other letters from reader of Murphy advertisements so tickled Frank Murphy, that he sent them to Rupert Casson to reproduce in an advertisement, so as to let the public share the joke, and incidentally, neatly illustrate Casson's statement that "we are not always on our dignity, and don't mind printing a joke at our own expense". (See page 44 advertisement headed "WHAT A LIFE!!!").

In Susan Briggs' recently published history and anthology of *THOSE RADIO TIMES* she notes (pages 40, 41) that "by the mid-1930's you could respond to Frank Murphy's personally signed advertisements", and she quotes his words:—"I shouldn't be appealing to you on moral grounds if I hadn't complete faith in Murphy sets" as being typical of Depression-style advertisements.

From this, you might think that such advertisements were typical of advertisers at this time, which they certainly were not. The first advertisement, headed unequivocally "*UNEMPLOYMENT*" appeared on April 7th 1933 in the *Radio Times* and leading national daily newspapers, and was a startling departure, in that it told people to "*Buy your Murphy sets now*", e.g. in the summer months. (See advertisement page 46). This would enable

UNEMPLOYMENT

WHAT'S TO DO ABOUT IT?
Let's put out the light and GO TO SLEEP!

WE'VE had a million debates, speeches and schemes on unemployment. Most of them good. And when they're done we don't seem much nearer a solution. I don't pretend to understand " world-causes " or currency problems or international debts, and I don't think many people do.

I *do* understand that at Murphy Radio we have an unemployment problem, and we know how it can be cured.

I am only one (and a small one at that) of thousands of employers, but I employ hundreds of workers. If a few thousand employers can each " clean his own doorstep," the unemployment problem will begin to look healthier. In this and my next advertisements I am setting about my own job. Two points want noting by you :—

1. In the end *you* (buyers) are the only people who make employment.
2. We are going to spend thousands of pounds on advertising at the time when it produces the least results.

If anyone says this is a stunt to sell Murphy sets, then he is an ass. WHY ARE WE DOING IT? BECAUSE IF YOU WILL *BUY* SETS EVENLY ALL THE YEAR ROUND, WE SHALL *MAKE* SETS EVENLY ALL THE YEAR ROUND, AND OUR WORKPEOPLE WILL HAVE WHAT, ABOVE ALL ELSE, IS THE FIRST RIGHT OF EVERY WORKER—*SECURITY OF EMPLOYMENT.*

I will just sum up what I am going to prove to you in the next few weeks.

1. You want wireless sets in the summer.
2. At least half of you *pay* for them all the year round even now, for you buy on Hire Purchase.
3. If you take delivery all the year round, you will get better and cheaper sets and give better and permanent employment.
4. There is no point in waiting if you want a Murphy set, for we will bring out no new models nor will we reduce our prices during 1933. Our new models are out *now.*

¶ Please take these announcements seriously, and if you agree with them—then ACT.

Frank Murphy
B.Sc., A.M.I.E.E., A.I.R.J.E., Chartered Elect. Engineer

A8 *For A.C. Electricity*
ALL ELECTRIC 8-VALVE SUPERHET
Single Tuning Control. Moving Coil Speaker. Illuminated Wavelength Dial. AUTOMATIC VOLUME CONTROL.
Walnut Cabinet. *Cash Price* **£24 - 0 - 0**

A4 *For A.C. Electricity* **D4** *For D.C. Electricity*
ALL ELECTRIC 4-VALVE SUPERHET
Single Tuning Control. Moving Coil Speaker. Illuminated Wavelength Dial. Walnut and Rosewood Cabinet.
Cash Price **£14 - 10 - 0**

B5 BATTERY 5-VALVE SUPERHET
Specifications as A4 and D4.
Cash Price **£14 - 10 - 0**

BUY YOUR MURPHY NOW!

I am not quite sure whether this set will be ready when this advertisement appears.

You can buy Murphy Sets only from appointed Murphy dealers. Any model on Hire Purchase.

MURPHY RADIO LTD., WELWYN GARDEN CITY, HERTS. TELEPHONE: WELWYN GARDEN 80 at least one manufacturer with a conscience to keep his workforce regularly employed and not throw them on the dole until the autumn when sales automatically soared. The advertisement was unusually long—the text took up three quarters of the page—and there were no concessions to "snappy sales talk". Moreover, at a time when most radio manufacturers cut their losses by minimal advertising, Murphy Radio ran an unprecedented advertising campaign right through the summer months, spending thousands of pounds, not on advertising new models (indeed, one of the main planks of the campaign was that since all the new Murphy models were already out for 1933, it made no sense to wait to buy one till the autumn), but on punching home the message that the buying public could individually contribute to a solution of the unemployment problem. The obvious cynical reply was countered by the statement, 'If anyone says this is a stunt to sell Murphy sets, then he is an ass'.

Every fortnight another advertisement appeared on this theme, until on June 23rd 1933, a typically candid statement appeared:—

"BETWEEN THE DEVIL AND THE DEEP BLUE SEA!
I feel you ought to know more about our new sets, but at the same time I want to keep your mind focussed on this summer unemployment business. The trouble is that a single advertisement is not big enough to do justice to both. So, in brief, this is the progress we've made on the unemployment side so far this sumnmer. Starting from January, the number of people employed in our works fell 20 per cent by the end of February. But by the beginning of June the whole of this 20 per cent had been taken on again. Don't think the job is finished, however. There is still June and July to be tackled. Now for the sets, and to begin with the *B5*. . . .

Here several paragraphs followed, from E. J. Power, Murphy Radio's Chief Engineer, telling the public about the new battery model (the *B5*), and why it had been decided to employ a particular kind of amplification method in it. This was clearly way beyond the comprehension of most *Radio Times* readers, and so Frank Murphy concedes:

"It is impossible, as I have said before, to give any clear idea of a wireless set in an advertisement. The obvious thing to do is to go and hear a *B5* or one of the other sets

for yourself at the nearest Murphy dealer. And then buy one NOW! So we get back to our unemployment theme after all—
BUY YOUR MURPHY NOW!"

One month later, the more familiar "Murphy style" was back. The Man with the Pipe asked his readers: "WHAT IS CHEAP?" (See advertisement, page   ) and once more, in half a dozen lines, punched home his "Value for Money" message.

That was in July. In August, the radio industry traditionally put its latest models on show at Radiolympia—but Murphy Radio sets were not there, much to the surprise of the public and the chagrin of many of the dealers. Frank Murphy had his reasons, and characteristically set them forth in an advertisement on September 1st, beginning:

"THERE'S NO PLACE LIKE HOME . . . One of the reasons why we were not at Olympia, is because sets cannot be demonstrated there. It seems rather like trying to sell spectacles without letting the customers look through them first. There is in my opinion only one satisfactory way of choosing your wireless set, and that is by testing several makes for yourself and picking the one you think is the best. If you are considering buying a set, you ought to do one of two things:—
1. Go to the man who runs a radio exhibition every day in the year. In other words, go to a good dealer who can show you several sets, and give you comparative demonstrations at any time.
2. Run your own radio exhibition in your own home. Personally, I should say that Number 2 is by far the most satisfactory method. I think that 'home' is the right place in which to decide to buy a wireless set. You are there at ease, able to make up your own mind in your own way. You and your family can see and hear the set where it is going to live.
"FRANK MURPHY"

In this advertisement there is a cunning emphasis on the idea of *'home'* as the proper setting for the wireless—one gets an instant picture of the typical family grouped cosily around their cherished set—an image redolent of the nineteen thirties. The television set of today has no such family associations; in fact, in many homes individual members of the family are watching different programmes on portable sets in separate rooms. The other phrase worth pointing out is the one about every buyer "making up his own mind in his own way"—not the sort of practice a high-pressure salesman would encourage, but very typical of Frank Murphy. He treated his potential customers as sensible human beings, perfectly capable of coming to satisfactory value-judgments, even though they might not have the technical knowledge of the radio expert.

As a result, the people who read his advertisements could not fail to be attracted by his unorthodox approach. There was his candid acknowledgment of problems not yet solved; his simple pride in the quality of his sets in reproduction and reliability; his trust in his dealers to give genuine advice and service; and his belief that the public consisted, not of ignorant fools to be exploited for profit, but of sensible men and women who could recognise a good thing when it was pointed out to them, and come to their own conclusions. His humorous, down-to-earth phrases delighted them, for they were just the kind of language used by ordinary people, and refreshingly different from the usual advertisers' jargon.

All this is reflected in what was perhaps the most famous one of all the early advertisements—the one which linked the new Murphy models for 1934 with the fact that Hilda Murphy had just won first prize with her entry of six new potatoes in the Welwyn Horticultural Show. The pun was too good to miss, and no doubt, not only Frank Murphy and Rupert Casson enjoyed the joke, but a great many other people too. (See advertisement, page 50)

But in September 1937 a change came over Murphy advertising. Frank Murphy had departed from Murphy Radio, and his face had disappeared from the advertisements, to be replaced by that of E. J. Power, the new Managing Director. Rupert Casson continued to write the advertisements, but there was a perceptible change of style—naturally enough, since E. J. Power, who was primarily an engineer rather than a communicator, did not share Frank's passion for broadcasting ideas and policies through the media of the press. The unpredictable innovator had given way to the more cautious but steady perfectionist. Murphy Radio settled down to become one of the leading radio manufacturers, and Frank Murphy set off on a new venture, which had no connection with radio.

WHAT IS CHEAP?

WE are often asked to produce a set at a much lower price, and we are told that "there is a huge market for a cheap set." I want to say quite clearly that whereas we don't object to huge sales, this is not the primary object of Murphy Radio.

Our real object is to make first-class radio sets. If we knew how to do it for five or six pounds we should certainly be doing it. But at present we don't.

I am told: "Think of the people who can only afford five or six pounds!" I *do* think of them, and I think they are the last people of all who can afford to buy a set with which they may soon be dissatisfied. It is pitiful to think of the hundreds of thousands of pounds which poor people have wasted on bad wireless sets, because they "couldn't afford" a good one. The rich can take a chance on "cheap" things. Those who have to count their shillings and their pennies can't!

It is my definite belief that Murphy sets are—in the true sense—the cheapest you can possibly buy.

Frank Murphy

All Murphy Sets obtainable on Hire Purchase Terms.

BUY YOUR MURPHY NOW!

A4 / **D4**
ALL-ELECTRIC 4-VALVE SUPERHETERODYNE

£14-10-0

A8 ALL-ELECTRIC 8-VALVE SUPERHETERODYNE

£24-0-0

B5 BATTERY 5-VALVE SUPERHETERODYNE

£14-10-0

1st PRIZE
for the New Murphies!

MRS. MURPHY (who is in charge of the Gardening Department of our household) has been awarded first prize for New Potatoes in the Welwyn Horticultural Show. We all feel that there is a deep significance in this, though we can't find out exactly what it is.

You will notice that there are six of these prizewinners.

Surely there is something in the fact that this year we have already produced five new models and the sixth, the Radiogram, is almost ready to be dug out (if that is the term).

Well it's not for me to labour the point, but I suggest that you call on one of the 836 Murphy Dealers and do a little judging yourself.

Mrs. Murphy is not the only one in the family who can win first prizes.

Frank Murphy

P.S.— Now I shall get into trouble for being undignified.

TABLE MODELS
1. A.24 for A.C. Mains.
2. D.24 for D.C. Mains.
3. B.24 Battery Model.

Cash Price **£14 . 10**

CONSOLE MODELS
4. A.24 for A.C. Mains.
5. D.24 for D.C. Mains.

Cash Price **£17**

6.★ RADIOGRAM — Coming out in August.

HIRE PURCHASE TERMS AVAILABLE.
These prices do not apply in I.F.S.

MURPHY

MURPHY RADIO LTD., WELWYN GARDEN CITY, HERTS.   PHONE WELWYN GARDEN 500

Chapter Six

PRODUCTION IN THE FIRST SIX YEARS (1930-36)

(For vintage radio buffs)

In 1936 the magazine *Cavalcade* printed an article on the remarkable success of Murphy Radio Limited, in achieving sales of 80,000 sets during 1935, having steadily increased from the modest start of less than a thousand sales in 1930. The reporter attributed this meteoric rise to a turnover of a million and a quarter pounds to the unusual outlook of its founder, Frank Murphy, who stated openly that he was not interested in making money, but in giving "value for money" to his customers.

There were two aspects to this policy. First, as described in Chapter 4, the most rigorous tests were applied to the design of each new model, to ensure quality and reliability; secondly, there was a fixed retail price which was calculated at a factor of 1.955 of the cost of actually producing the set, in terms of materials, labour and direct overheads. Just less than half the retail price had to cover the cost of research, advertising and selling, as well as a strictly controlled profit margin for the retailer, and the manufacturer's own profit. In addition, Murphy Radio offered a full year's guarantee to repair or replace any faulty parts, free of charge.

When the first Murphy set appeared on the market in the summer of 1930, it was priced at seventeen guineas (£17 17s. 0d.). This was the four-valve, battery-operated *B4*. Its main competitors were a five-valve portable by Pye at 19 guineas, and their "twin-triple" battery portable at 22 guineas; the Philips 25/22 portable at £27 10s. 0d.; the Ekco 3-valve mains set with separate speaker at £22 10s. 0d.; the Kolster-Brandes model 103 4-valve battery set at 19 guineas, and the Marconiphone portable model 55 at 18 guineas.

The Murphy *B4* was cheaper than all of these, even at the time of its introduction, and by February 1932 an improved *B4* was selling even more cheaply at 15 guineas. The price was cut yet again in January 1933, to £12 10s. 0d. so no wonder Murphy owners felt they were getting value for money. Incidentally, one of the first *B4* models was bought by Clifford Stephenson of Huddersfield at the start of his career as an appointed Murphy Dealer. Years later, he bought it back from the man to whom he had first sold it, and presented it to his local museum as an example of fine workmanship.

The *B4* owed its design to Frank Murphy and E. J. Power, as explained in a previous chapter, but once production began in earnest and the task of selecting and briefing the special Murphy dealers became increasingly absorbing of their time, it was clear that others must be recruited to do the future design work. Frank Murphy therefore approached his former colleague Professor Mallett of Imperial College, London, and asked him to recommend his most brilliant student in electrical engineering. It was in this way that Dr. (now Professor) Robert C. G. Williams came to be invited to head Murphy Radio's research team, and under his inspirational direction the team produced a series of models which put Murphy Radio in the front rank of the radio industry.

The first breakthrough came in August 1931, when by tradition all the radio manufacturers brought out their new sets to capture the pre-Christmas trade. Murphy Radio, a virtually unknown company, introduced their first mains-operated table model, the *A3,* and scored a great triumph when visitors to the Exhibition at Radiolympia voted it the Outstanding Model of the Year. The price of the *A3* was 19 guineas, and it continued in production with improvements as the *A3A* through 1932, when the price was reduced dramatically to £13 10s. 0d. It was the first Murphy set to be housed in a cabinet expressly designed for it by the famous furniture firm of Gordon Russell Limited (see illustration on page 64). How this came about is another story which is related a little later.

For the moment, however, we return to technical details. It was in 1932 that Murphy Radio produced their first major technical triumph—the introduction of the *A8,* an eight-valve super heterodyne receiver with automatic volume control—the first of its kind, and at the incredibly low price of £24 0s. 0d. The *A8* was the first British attempt to provide a receiver which, while

"VALUE FOR MONEY" graphically illustrated . . .

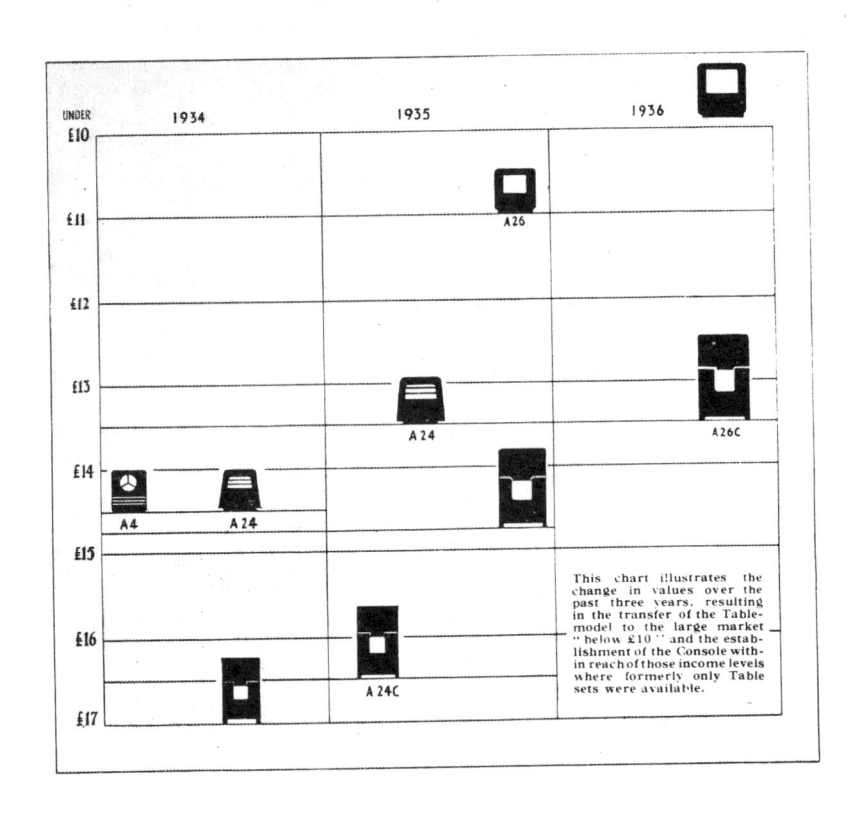

This chart illustrates the change in values over the past three years, resulting in the transfer of the Table-model to the large market "below £10" and the establishment of the Console within reach of those income levels where formerly only Table sets were available.

simple to operate, would receive distinctly about fifty foreign stations, and would maintain them at a constant signal level. (See illustration, page 50) Only a superhet circuit could economically provide the range and selectivity, and only automatic volume control could ensure the constant sound level.

There was quite a gap between the *A8*, Murphy's luxury model at £24, and the *A3A* at £13 10s. 0d., both in price and performance. This was filled early in 1933 by the *A4*, a four-valve superhet receiver, with that circuit's advantages of great selectivity and long range, yet with single knob control. Both A.C. and D.C. models retailed at £14 10s. 0d., and at the same price there was a new battery superhet—the *B5*.

In 1934 these designs were further developed in the *A24*, *D24* and *B24*, which, while still priced at £14 10s. 0d., were now all fitted with automatic volume control, with three times the 'gain' of the previous year's models. (See illustration, page 70).

The next year, 1935, saw the price levels come down dramatically, and the value for money concept became even clearer when the public were able to buy the Murphy table models *A26*, *D26* and *B26* with automatic volume control and refined circuiting for only £11. (Even this achievement was surpassed in 1936 when the table models came down to £9 17s. 6d. and the company produced its first really inexpensive battery model, the *B23*, to retail at £6 7s. 6d.)

The inclusion of D.C. mains models during these years was necessitated by the existence of considerable pockets of direct current electricity supply in many urban areas, at a time when the standard electricity grid supply of 230 volts or 50 cycles had not penetrated to all corners of the United Kingdom.

Meanwhile the Murphy design team were intent on yet another breakthrough. Everyone was delighted with the remarkable selectivity and range of the inexpensive superhet table receivers, but there was increasingly a popular demand for better audio reproduction or, as the Murphy advertisements put it, "beauty of tone". So in 1934 the engineers triumphantly produced the first Murphy console (floor-standing model)—the *A24C*—at a price of £17.

The *A24C* had several interesting new features. A much larger loudspeaker ensured both low natural resonance and exceptionally brilliant "attack" in the treble, of a kind not previously known in a commercially-produced set. Next, the console cabinet was designed with narrow back-to-front dimensions, so limiting "boomy" box resonances; and this problem was still further decreased by leaving the back and the bottom of the cabinet open, and enclosing only the receiver chassis. Vibration from the sides of the cabinet ("wood resonance") was obviated by the use of thick wood for the cabinet and by lining it with a padding of sound-deadening material.

The price of the first Murphy console model was only £17. Yet in 1935 its successor, the *A26C,* was selling at £14 15s. 0d. This meant that people who could afford to buy a Murphy table model in 1934, could now buy a console model in 1935, for almost the same money, but with a very great improvement in quality of reproduction. (See illustration of the *A24C* on fig. 18).

Next on the agenda for the Murphy engineers was a radio-gramophone, and the first one, the *A24RG,* came out in 1934, the same year as the first console model. In sharp contrast to the typical manufacturer's fretwood box on mock-Jacobean legs, the Murphy radiogram was housed in a simple console-type cabinet of Indian laurelwood. The gramophone turntable equipment incorporated an automatic switch and brake developed entirely in the Murphy laboratory. The sound reproduction of both radio and gramophone was outstanding for its time, yet the price was only £24. This, it is interesting to note, was the price two years before of the first Murphy superhet!

In 1935 Murphy Radio made yet another contribution to the development of the radio receiver, namely, automatic tuning correction. This ingenious device enabled even relatively unskilled persons ("such as womenfolk, for example") to obtain a desired station dead in tune. It worked by the interaction of two valves, a double diode and an H.F. pentode, which occurred when the tuning knob was released; but while the knob was being rotated, a mechanical clutch operated a switch throwing the automatic tuning device out of circuit until the knob was again at rest, when the automatic frequency correction circuits pulled the station accurately into tune.

Another innovation was a noise-suppression system, to reduce the level of background hiss experienced when tuning between stations. A switch at the back of the receiver allowed the user to set the tone to either "mellow", or "brilliant" (i.e. with greater top clarity).

Both these new devices were incorporated in the *A28C* and *D28C* console models which like the "26" series of the previous year had image frequency suppression, automatic volume control, large loudspeaker, and padded speaker compartment unenclosed at the rear. The price for both console models was £21 15s. 0d.

As might be expected, the 1935 radiogram was also fitted with automatic tuning correction and the noise-suppression system. But what was unexpected was its shape. Not everyone approved of its assymetrical appearance, with the gramophone lid covering one half of the top of the square-shaped console, while the other half of the top disclosed the tuning scale for the receiver (see illustration on fig. 19). Unusually, too, it was (at £33 10s. 0d.) nearly a third dearer than its predecessor, but then it incorporated the new technical developments in the radio receiver, which had an eight-valve circuit instead of a four-valve one, and two L.F. stages instead of one.

The hectic pace of the first five years and consequent feeling of exhilaration could not be expected to continue indefinitely. By 1936 there was a feeling that the radio receiver had reached a comparatively stable stage of development, and Chief Engineer E. J. Power summed it up by saying, "The wireless receiver has now reached an evolutionary, rather than revolutionary stage". After this, it was likely that the company would be offering increasing value for money not so much by way of technical innovations, but by offering models similar in design to the previous year, but at keener prices.

Nevertheless, one new type of receiver was produced in 1936—the *B23* battery portable (already mentioned), which gave performance comparable with that of a superhet without the additional cost of superhet circuitry, and so could reach a new section of the public at only £6 7s. 6d.

Murphy Radio's competitors, however, were quick to take advantage of the lull in technical advances from Murphy, as the dealers were ready to point out, and in particular they could not see why Murphy Radio did not mark their tuning dials with station names, as other manufacturers were doing. (Some of their comments are given in the chapter on the *Murphy News*.)

But in 1937, Murphy Radio once more announced a "first" for

the company—the tuning dials were to be not only marked with station names, but these names were given *in alphabetical order*—a technical triumph for the men in the Murphy design team.

There were three new models to rejoice the hearts of the dealers; a superhet table model, the *A32*, to sell at only £8 5s. 0d.; the *A36*, a short-wave receiver giving world-wide reception, at £15 10s. 0d.; and a completely re-designed and greatly improved console (*A38C*). Finally, looking well ahead, the company announced that they were going into production with their *first television receiver* (the *A42V*), and revealed that they had been asked to provide one of these for the special Television Exhibition at the Science Museum. It is pleasant to record that many years later an *A42V* still in working order was presented for permanent exhibition to the Science Museum by Mr. Roy Drew of Croydon, a former employee of the Rank Organisation.

As we have seen, in a span of only six years Murphy Radio had achieved remarkable success in cornering a significant part of the expanding market for radio receivers in Britain. The "value for money" in engineering terms was evident enough. Probably even more outstanding was the distinctive style of the Murphy cabinets, which even today can be appreciated as fine pieces of furniture in their own right. It was the first time that anyone using mass-production techniques had been able to bring the qualities of honest design and skilled craftmanship, previously available only to the rich, into the homes of ordinary men and women.

How this came about is a typical Frank Murphy story. During the long months of testing and research which culminated in the production of the first Murphy set—the battery-operated *B4* portable—he had little time to spend on the outward appearance of the set, beyond determining that the set must be a pleasure, not only to listen to, but also to look at and to handle.

He took the traditional path of contacting an established firm with experience of producing cabinets for the radio trade, and gave them firm instructions which were somewhat unusual. The cabinet for the *B4* was to be as simple as possible, with no frills or fancy fretwork, in a wood which blended with living-room furniture of that period—which turned out to be dark-stained and highly polished walnut. The tuning scale had to be easily visible; so the top of the cabinet, instead of being entirely level, was angled to take the tuning knobs and scale window, while the operator's hands rested on the level section. Concealed handgrips were insisted on, to allow the "portable" (which weighed 32lbs—quite reasonable in those days!) to be moved without difficulty. Above all, the cabinet had to be strong, as it would need to house both high tension and low tension batteries. (It is said that the early Murphy representatives liked to demonstrate just how strong the cabinet was, by deliberately dropping the set on the floor; but this is probably apocryphal.) The result was a simple, practical design, but hardly an object of beauty.

Frank Murphy had no doubt that the cabinet for the next set—the table model *A3* would need to be really good to satisfy him, yet it was obvious that the regular suppliers of radio cabinets were not going to be able to design anything better. If he wanted a piece of fine furniture, the logical step was to seek out a designer capable of producing it. So he went to see two people who he thought would be able to advise him; the Editor of the *Cabinet-Maker* and the Principal of the London County Council Technical Institute at Shoreditch. Each of them recommended "a man called Gordon Russell" as someone particularly interested in good design. It seemed a worthwhile possibility for investigation. Frank Murphy rang up Gordon Russell, then in his furniture workshop in the beautiful Cotswold village of Broadway, and asked him if he was interested in making radio cabinets. Russell, quite taken aback, said he knew nothing about it. "That doesn't matter", was the reply, "If you're interested, we'll come down tomorrow afternoon."

In his autobiography, *Designer's Trade*, Sir Gordon Russell has left a hilarious account of his first encounter with Frank Murphy and Ted Power. Sadly, after a long struggle with a disabling illness, Sir Gordon died in 1980. Many tributes were paid to his pioneering work, both in the field of design and in bringing about the recognition by industry of the importance of the designer. His son Michael has kindly given me permission to quote from Sir Gordon's book:—

> "So down to Broadway they came—'they' being Frank Murphy and his partner, Ted Power, wearing old mackintoshes and cloth caps and looking as if they worked round the clock. Dick (Russell's brother, later Professor R. D. Russell)

and I met them, but no time was spent in formalities. Murphy launched into a description of his project in a spate of words... 'Look at this', he said to us, producing a portable cabinet, 'It's just a box. No ideas. Ted and I have spent many hours trying to find how we can keep these ugly knobs out of sight without making them inaccessible but we haven't got anywhere. We're at a dead end. What's the next move? The matter is desperately urgent. We've been working on the set for a year, we're starting to tool up, and sets must be in dealers' shops well ahead of Christmas. The portable will have to go as it is, there's no help for it, but the next model just must have a better cabinet; the problem is, how?' He stopped talking, and they both looked at Dick and me. I realized how deadly earnest he was: the whole future of both of them seemed, in their eyes, to hang in the balance. I felt that the only thing to do was to put all my cards on the table, so I said: 'I'm not an engineer and I'm not a wireless fan, I haven't got a set—I listen in only occasionally at my father's house—and I must admit I've never studied the design of such cabinets, although I have occasionally looked at them. And I'll tell you what I think is the trouble. The whole approach to the problem is wrong. This is what I imagine happens. The set-maker sends a blue-print to a cabinet-maker and asks for a few sketches. The cabinet-maker does no real research into requirements. His draughtsman—who is not a designer—knocks off a few sketches, then some samples are made and sent in and one is chosen. The set-maker or the cabinet-maker buys some cloth for the loud-speaker opening, asks a printer to set out the names for the stations and buys some plastic knobs from a catalogue, or perhaps he first sends the size to the manufacturer and leaves the choice to him. No one competent designer has been responsible for all the visible parts of the set and has welded them into a complete whole, and until some pretty fundamental research is done and the whole business is approached in a new way, I can't see that you can expect much advance. Now is that a true picture of what happens?'

'Where did you get your information?' said one of them. 'That's just about what does happen.'

'From the set', we said. 'Consider the problem of the knobs, which are not for pulling out a drawer—although by their shape you might think they were. By a careful study of the requirements you could make them much more pleasant to handle and to look at without costing any more. They should be designed as an essential part of the whole job. If they are good to look at and well placed, they will enhance the cabinet and it will never again occur to you that it would be desirable to hide them. The design of the cabinet ought to start when you design the set, not as an after-thought.'

They sat back in their chairs and looked at us in astonishment. That such simple facts should have escaped highly intelligent men shows how low design had fallen. They said they hadn't felt so hopeful for months and Dick and I promised to go down to their works and discuss the matter further on the spot. They had all the engineer's confidence that a thing could be made once a specification was laid down, but to me cabinet production presented a considerable problem. We went down to Welwyn Garden City, where they had a small factory, and it was obvious that they had given the same concentrated attention to every side of their job. I loved the way they took nothing for granted. They wanted to know the why and the how. Frequently C. R. Casson, their advertising agent, was present at these discussions. He had a humorous face and liked to pretend he was just an ordinary member of the public, testing their odd theories. At a pause in the spate of words he would sit back in his chair and say: 'I'm Billie Muggins, so you must explain what the hell this new idea is! There's no doubt I'm dull-witted, so make it simple.' 'Making Wireless Simple' was Murphy's slogan, printed on every carton.

To my surprise, the question of public acceptability never entered the discussion at this stage: they wanted to find the best possible solution, to tell the public about it and then they felt the public would accept it. They were pioneers, with a zest for the job, and that link held us together through several very tricky years. I never remember them saying, 'Oh, you can't do that—it won't sell'—like so many people in the furniture trade. I think that to intelligent hardboiled engineers there seemed something a bit odd about wrapping an imitation Queen Anne case around such a precise and complicated piece of machinery. This was an advantage. The furniture maker was not brought quite so suddenly against his own age: he sold his wardrobes neat whereas people bought a radio set on its performance and put up with the cabinet.

We began to state the problem on paper, always a useful initial stage to finding a solution. The engineers, accustomed to working in thousandths of an inch, laughed at our idea of a tolerance: a sixteenth of an inch was a crevasse to them! They said we must learn to be accurate, wood or no wood. All measurements had to be from the inside, whereas we were accustomed to working to external ones, and were tied by a number of points where the set had to register. There were problems of supply, of sound, of handling on their benches, of talking to dealers, of packing, and so on. I noticed that they were just as thorough over the last item: they worked out their own cardboard carton with special loose lining and then sent several complete sets all round the British Isles to see what happened.

Never a day went by without their having some criticism to hurl at us. It might be a minor detail, such as a small alteration in the fixing of the silk, or it might be a bolt out of the blue. I remember one evening at home when, just as I was about to sit down to a delectable omelette, the telephone rang and on taking up the instrument I heard Murphy's monotonous voice—he never said 'Good Morning', 'Good Evening' or even 'Blast you'—'I've been thinking about this cabinet problem and I want to ask you this: why do we need a cabinet at all! Aren't we just accepting an out-of-date solution! I think we've got to get right down to first principles on this job. Perhaps it isn't sense to ask a cabinet-maker whether we need a cabinet, but I trust you to ponder this carefully. This isn't just a stunt . . .' and so on, like water pouring over a mill-dam, for more than a quarter of an hour. I thought wistfully of my omelette, and at the first sign of a pause I said: 'Well, Murphy, my first reaction is that no woman will dust all your bits and pieces when she can put a duster over our cabinet in a twentieth of the time. There may be other points. Certainly I'll think about it. Good night'. And as I put the receiver back I said a little prayer for all poor souls who are subjected to the third degree. On another occasion he propounded the theory that nothing should cost more per pound than beefsteak. He said a Ford car didn't, but our cabinet did. 'Well', said I, 'do you suppose people would accept beefsteak in lieu of cabinets?' I forget how the argument ended.

Intensive research into the problem led to a tentative solution finding its way on to paper and when this was approved a sample cabinet was made. It was unlike any former radio cabinet so far as I know (see page 64). The carcass was of solid walnut with two veneered plywood panels, and there was a loudspeaker opening with a grid in front of the grey silk. The control knobs were not good, but they were much better than in the first model; and owing to the engineers' layout, they were in positions which made it somewhat difficult to harmonise them with the cabinet. This demonstrated clearly that to attain a really good design one would have to work in collaboration with the engineers at the earliest stage, that is, before tooling up, when such points were often to some extent fluid. Once again, may I repeat that good industrial design goes down to the roots—it is never something added at the end. I remember Lethaby's saying: 'Art is not a sauce added to ordinary cooking. It is the cooking itself, if it is good'. Both Murphy and Power were pleased with the experiment and the time came to show it to the dealers as the new season's model.

I think Murphy and I went to this jamboree. There was, if I remember rightly, a lunch beforehand of the kind that jaded hotel keepers imagine that still more jaded businessmen are bound to like. After this ordeal by mastication the cabinet was produced. The dealers stood up and gazed at it in rapt astonishment and alarm. Then some wit, noticing the grid, christened it the Pentonville and the rest roared with laughter. Murphy was magnificent. He knew nothing of design himself, but he had made up his mind that Dick and I did, and that he was going to back us. The reception didn't rattle him a jot. He stuffed his pipe in his mouth and dealt with various questions with perfect good humour. Then someone said:'It's a good set, but no one will buy it in a cabinet like that. Why don't you get some bloke who knows

RADIO TIMES — May 29, 1931

MAKING · WIRELESS · SIMPLE

**MURPHY TYPE A.3
3-VALVE ALL-MAINS RECEIVER**

SINGLE TUNING CONTROL DIAL MARKED IN WAVELENGTHS

Two other controls — Wavelength Switch and Volume Control.

Tuning Dial illuminated.

Band pass filter as first tuned circuit.

Screened grid, Detector and Pentode valves.

A test yielded reception of good programme value on outside aerial, and 15 stations on 12 foot inside aerial, all stations at perfectly clear and loud not included in the report.

Self-contained moving coil loudspeaker with energised field coil.

Quality of reproduction excellent with ample top and bottom register.

Quality substantially maintained from very reasonable to considerable volume.

Walnut cabinet of very distinguished appearance by one of the leading designers of the Cabinet Industry.

Polished black ebonite escutcheon plate.

CASH PRICE
COMPLETE · · **19 Guineas**

Hire Purchase: £5 deposit and 12 monthly payments of £1/5/-. Or £3 deposit and 12 monthly payments of £1/10/2d.

I HAVE just prepared a booklet entitled "Making Wireless Simple" which gives you a simple explanation of how loudspeaker works. It also contains brief particulars of Murphy sets. I shall be pleased to send you a free copy on application.

T HIS is my new set. It is certainly outstanding in design, isn't it?
— Well it's just as outstanding in construction.
I am fully convinced that you and your friends will buy this set, because in *your* opinion it is the finest you can buy at anything like the price.
My factory staff, sales representatives and all Murphy Dealers are full of enthusiasm about it. They are easily as keen as I am.
I do feel that the Murphy All-mains 3 is a great step forward in my policy of making wireless simple.

Ask your dealer to demonstrate one to you.

Frank Murphy
B.Sc., A.M.I.E.E., A.I.R.E.E., Chartered Elect. Engineer.

MURPHY RADIO, LTD., WELWYN GARDEN CITY, HERTS. WELWYN GARDEN 331.

MURPHY RADIO

ALL ADVERTISEMENTS FOR "RADIO TIMES" SHOULD BE ADDRESSED TO ADVERTISEMENT MANAGER, B.B.C., 8-10, SOUTHAMPTON STREET, STRAND, LONDON, W.C.2. TELEPHONE: TEMPLE BAR 8400.

the job to design it? There must be plenty about. Why, this bright-eyed guy you've picked on doesn't even know the right colour for the silk and in my opinion it's wasting time to have people like that.' A murmur of approval ran round the room. Murphy asked me to deal with this, so, in my most ingenuous manner I asked: 'Could you tell me the right colour for the silk?'

'Why', said the dealer, now quite certain that he was dealing with a fool, 'it's a kind of coppery colour'.

'What makes you say that?' I said.

Hardly believing his ears and astonished that such greenhorns should roam about unattended, he replied testily: 'Why, you've only got to look in the window of any radio shop and you'll see that all the sets have a copper-coloured silk.'

'Yes', I said, 'that's true, but it does not prove that the colour is right. Anyone who knows anything about colour knows that a hot coppery colour does not go at all well with walnut, which is used for these cabinets. A cool contrasting colour such as grey is much more suitable and that's why we've used it. It will never be Murphy Radio's policy to go through the gate just because it's been opened for the sheep. They have original ideas on a good many sides of their business, as you know, so you may as well realize there's going to be a Murphy Style in cabinets. The only reason for the coppery coloured silks is that a set which sold exceptionally well three years ago had it for the first time. Isn't that true?' A few of the dealers approved the suggestion, and the brickbats slowed up a bit.

I attended meetings of this sort over several years with Frank Murphy and I learned many valuable lessons at them. Perhaps the most important was the one every politician has to take to heart: never, under any conditions, let the hecklers rattle you. So long as you bear this in mind and know clearly what you want to do yourself, you have most of the cards, for few hecklers have a constructive policy. Some meetings were pretty noisy, but the dealers were not slow to see that we meant what we said and gradually the supporters of our cabinets increased. The astonishing thing was that in a few years' time the dealers, when shown the new models, often said: 'Nothing like as good as last year's *A3*. That *was* a cabinet, that was!'—blissfully unconscious of the fact that they hadn't a good word for the *A3* when it was introduced! But for Murphy's personal interest and championship, the whole idea could hardly have survived the teething stage. Ted Power was just as interested and perhaps a more able critic, but he was tied up on production.

The variety of models to be made increased each year both in number and, with many of them, in size. Production became a very big job—40,000 of one model would have seemed fantastic to us but a few years before. We built a small factory in Broadway but it didn't prove large enough and in 1934 we moved this part of our work to a new factory at Park Royal, near Willesden, where we also did a good deal of work for other firms, such as Pye, Ultra, Ekco, etc. Looking back, there can be no doubt whatever that *this experiment of Murphy's powerfully affected the design of all radio cabinets made in England and I would say that the standard of the best in this country between 1930 and 1939 was as high as anywhere in the world.* Such were the spectacular results that were possible in a country where it was exceptional to find any object whose form, colour, texture and so on had been the subject of careful research by a skilled industrial designer—in this case, my brother Dick, who designed practically the entire range. We were able to prove that a very high standard of design, material and workmanship could be obtained in wood by mass-production methods, and that when such large numbers were involved, the cost of employing first-rate industrial designers added hardly anything to the price: in fact, it often reduced it by a better planned approach to the problem"

To Gordon and Dick Russell it soon became clear that designing for mass-production was bound to entail a different approach from that of the skilled individual craftsman. To be successful, mass-production had to be a team job, with the cabinet designer involved right from the beginning with the engineers, the press tool foreman, the weavers, the plastics manufacturers, the men at the mill and the assembly foremen; and then later with the advertising agents, the sales managers and the dealers. It was inevitable as a result of this need for close collaboration that Dick Russell should join Murphy Radio as their official cabinet designer. During subsequent years he produced a series of fine cabinet designs of consistently high quality, and was thus to introduce practical examples of good craftmanship into households that in other circumstances would never have acquired them.

It was the Murphy dealers who provided the first public reaction to the designs; and as we have seen from Sir Gordon Russell's account of their shocked reception of the first Russell cabinet, they were going to be very hard to convince. Gradually, however, they began to realise that the distinctive appearance of the $A3$, so different from their preconception of what a "wireless set" ought to look like, was in fact a strong selling point; they saw that the unusual cabinet was yet another demonstration of the superiority of Murphy sets, because more care and thought had gone into its design.

After a few months, dealers had overcome their initial aversion, had accepted the new cabinet and even come to be extremely proud of it; indeed, many dealers hoisted giant versions of the set on the tops of their delivery vans, and won prizes (and valuable publicity) by entering them in local carnivals.

Nicknames abounded, of course, and after the "Pentonville Three'"-the $A3$, with the speaker covered by a strongly marked grille—came the "Isle of Man Four"—a reference to the three-pointed metal star on the $A4$ cabinet of 1933. The first console model, the $A24C$ was even affectionately christened "Trouser-Legs", because the lower half and plinth were made of walnut, with an upper section of macassar ebony. I am not sure what the Russells thought of these nicknames, but they certainly appealed to Frank Murphy's sense of humour.

Dick Russell's first two cabinets, the $A3$ and $A4$, were accepted fairly readily; the wood used was walnut, which could merge discreetly into living-rooms furnished in many different ways. In 1934, however, he produced a sensational design for the new $A24$. Though the basic structure was still walnut, it was not treated with dark stain, but left in natural colour with a protective polish. Next, there was a front control panel of bird's-eye maple, much lighter in colour than the surrounding top and sides of walnut. Further, across the speaker opening were placed three bars of beech, an even lighter wood, and each bar was outlined with a scarlet line. It was a bold and striking design which certainly did not merge discreetly into domestic backgrounds. As Dick Russell himself observed: "The cabinet is either thoroughly liked or thoroughly disliked, and I think this is an encouraging sign. It is more or less easy to design a perfectly genteel and non-committal cabinet, about which little is said either way; but I am sure that such a cabinet is not likely to be so successful as one which gives rise to some heated argument."

Many of the dealers were so alarmed that they called on Murphy Radio to withdraw the *A24* cabinet for re-design, or at least to stain the light wood a darker colour. Such was the strength of feeling that an article appeared in the *Murphy News* (June 16th 1934) with an illustration showing an *A24* cabinet with half its maple apron stained, and half left unstained. Frank Murphy was reported as discussing the whole subject with Dick Russell, and asking him, "Why use wood at all? Why not use metal or bakelite?" To which the reply was, that wood was used because in the hands of skilled designers and craftsmen it could be selected to achieve very beautiful effects. "Why, then, *is* wood varnished and/or stained?" continued Frank Murphy. "Varnish is to protect the surface from damage, like plating on a chassis. Stain is purely for colouring—and, of course, it is a bad reason", said Dick Russell. "If you select a certain wood because of its inherent beauty, the attractiveness of its colour and grain, why go to a lot of trouble and expense to change it?"

Both of them knew, of course, that poor quality wood was often disguised by staining to achieve a spurious effect of expensive material, but the final product could hardly be described as genuine.

The argument was concluded by Frank Murphy observing: "It's up to the dealers and the public to decide, therefore, whether they want beauty in wood or beauty in stains—and when it's put like that, I know what most of them will answer!"

It was typical of Frank Murphy that, far from concealing the mixed reception the new set was getting, he chose to take a full page advertisement in the *Radio Times* to tell the public about it, and to explain why he did not propose to change the cabinet design (see page 70). Under the headline " . . . the longer you look at it the more you like it!" he took the opportunity to re-state his business philosophy.

"First of all, we are not making either wireless sets or cabinets with the main idea of selling as many as possible. We are trying to make them as *good* as we can."

Meanwhile one dealer at least decided to show the public that the *A24* looked perfectly at home in the living room. He asked Dick Russell to arrange some typical Gordon Russell furniture—chair, bookshelves, standard lamp, occasional tables—in his North West London showroom window, and among these were placed three *A24*s. This was Duncan Watson of Hampstead. Other dealers were urged to follow suit, even if they could not obtain Gordon Russell furniture, by showing the receiver in a room setting.

The ultimate proof of acceptance by the dealer and the public came in June 1934, when "Jane" in the *Daily Mirror* cartoon wailed, "This 24-hour clock on the wireless is simply ruining my wallpaper!" The artist showed her scribbling the old and new times of her favourite programmes (Henry Hall, Ambrose), on the wall behind her set—unmistakably an *A4*!

After this, the cabinet designs for the new models were looked forward to by the dealer almost as eagerly as the sets themselves, and letters of praise or disapproval were regularly printed in the *Murphy News*.

There is little doubt that, just as the sound reproduction of Murphy models was an enormous improvement on the customary "plummy" tones of conventional "wireless sets", so the design of the cabinets, at first so startling, came to be keenly appreciated by many people who had never before looked critically at their everyday environment.

Among those thus affected was Frank Murphy himself, as well as his wife and family. After a visit to the Broadway showrooms of Gordon Russell, Frank and Hilda decided to replace their existing furniture, and to re-decorate and completely re-furnish their house at Ludwick Corner (then on the edge of Welwyn Garden City) to which they had moved in 1936. Some of this fine furniture is still owned by the family.

There was another long-term result of the original decision to consult the Russells on cabinet design. Having grown accustomed to the sight and feel of fine furniture, Frank Murphy gradually came to the conclusion that there was no good reason

NOW THE D24 FOR USE ON D.C. MAINS

ALL ELECTRIC SUPERHETERODYNE RECEIVER. Totally enclosed moving coil speaker. Cabinet finished in Walnut inlaid with beautifully-grained Birdseye Maple. Single tuning control. Illuminated wavelength dial. Automatic volume control. Fitted with gramophone jack and extra loud-speaker sockets. Will receive British and many Continental stations and has very beautiful reproduction.

£14.10⁵

Cash Price

As advertised on A.C. Mains. Specification and price as above.

Hire Purchase Terms Available.

This price is not applicable in I.F.S.

"...the longer you look at it the more you like it!"

THERE is a big disagreement of opinion about the A.24 Cabinet design, and quite a lot of people think we ought to change it and give you "something that everybody will like."

Well, I'm not going to change it, but I do think you are entitled to know why. First of all, we are not making either wireless sets or cabinets with the main idea of selling as many as possible. We are trying to make them as good as we can. So far as the set is concerned, it is a plain fact that Murphy technique is now definitely leading the industry. And it does so chiefly because our Chief Engineer—Mr. Power—and his staff are trying to make fine sets, not "sellable" sets.

But when we come to beauty of design we are on very difficult and dangerous ground. Most of us will admit that we cannot lay down laws on how to achieve beauty. *I* certainly can't. So we did the obvious thing. We put our cabinet design in the hands of Gordon Russell, who give their life to trying to find and build beauty in the design of furniture. And if you could see their works in Worcestershire, or their showrooms in London, you would agree that they succeed.

Beauty in machine-made furniture is so rare that most of us do not recognise it when we see it. And many of us are not even willing to try! Meanwhile, in 3 weeks thousands of people have bought the A.24 and they know that what I say is true.

The longer you look at it the more you like it.

Frank Murphy

MURPHY

MURPHY RADIO LIMITED · WELWYN GARDEN CITY · HERTS · TELEPHONE: WELWYN GARDEN

MURPHY
NEWS
Saturday, June 2, 1934
JANE'S JOURNAL—Or the Diary of a Bright Young Thing - - Time Leaves Its M
THE DAILY MIRROR
I NEVER WAS GOOD AT MENTAL ARITHMETIC
AND THIS 24 HOUR CLOCK ON THE WIRELESS
IS SIMPLY RUINING
MY WALLPAPER!
(Copyright in all countries.)
WOMAN SECRETARY
A Compliment Implied—Accepted
VOL · 2 · No. 12

why the majority of the British public should not have the same pleasure. It was something he was to remember when the time came to leave Murphy Radio and to start another business.

Chapter Seven

"MURPHY MADNESS"

Much Madness is divinest Sense
To a discerning Eye . . ." *Emily Dickinson*

One of the original Murphy dealers recently wrote, "possibly Frank Murphy's greatest memorial is in the fact that a fair number of his 'disciples' are still in existence, and are still endeavouring to apply knowledge with integrity in the service of society—which was his creed".

Who these disciples were, and how they became a "band of brothers" is the subject of this chapter.

It will be remembered that Frank Murphy has resolved from the outset that the people who bought Murphy sets were to be given the utmost value for money, which led him to the decision to limit the selling outlets to one very carefully selected dealer per shopping centre. He gave these dealers powerful backing in his national advertisements, and advice and support through his representatives and the medium of the *Murphy News*. He encouraged them to visit the works at Welwyn Garden City. So many of them did so that a special Visitor's Department had to be created. Above all, he loved meeting them and talking to them, just as he loved talking to people in the factory and laboratories at Welwyn. He was just as happy to drop in informally for a chat on his way through a town as to face a critical, even hostile, assembly of them at a regional Dealers' Meeting—and to disarm them with his simple enthusiasm and puckish humour. The dealers soon learnt that whatever Frank Murphy said he would stand by, and while he pulled no punches if he thought individual dealers were slack, stupid or, worst of all, dishonest, as a body he defended them against all comers with the slogan, "Murphy Dealers are people you can trust".

The result was that in turn the dealers trusted him, and found

MURPHY MADNESS!

During the past three years a serious epidemic known as "Murphy Madness" has been breaking out amongst my dealers. Some get it in a very acute form. They throw every other set out of the shop and devote the whole of their time and energy to selling Murphy sets.

Others get it in a milder form, and although they sell all kinds of sets, they're happiest when they're obtaining Murphy "converts."

I think I am right in saying that not one Murphy dealer has escaped this epidemic in some form or another.

Here's the moral.

A dealer, if he wants to continue in business, cannot afford to give his customers anything else but a square deal. If more than a thousand picked dealers, all with technical knowledge, are prepared to "put their shirt" on Murphy Radio, you can be sure of one thing:

They all believe that by selling Murphy sets, they are giving their customers the best deal they can. If they got more by selling Murphy sets, *that* might explain it, but they don't. In fact in many cases they get less.

YOU CAN BUY MURPHY SETS ONLY FROM APPOINTED MURPHY DEALERS

Frank Murphy

Any model on hire purchase. *These prices do not apply in I.F.S.*

A4 *For A.C. Electricity.* **D4** *For D.C. Electricity.*
ALL - ELECTRIC 4 - VALVE SUPERHET
Single Tuning Control. Moving Coil Speaker. Illuminated wave-length dial. Walnut Cabinet inlaid with Rosewood.

£14-10-0 *Cash Price*

B5 BATTERY 5-VALVE SUPERHET
Class B Amplification
Single Tuning Control, Moving Coil Speaker. Illuminated wave-length dial. Walnut Cabinet inlaid with Rosewood.

£14-10-0 *Cash Price*

Models A4, D4, B5

A8 ALL - ELECTRIC 8 - VALVE SUPERHET
For A.C. Electricity.
Single Tuning Control. Moving Coil Speaker. Illuminated wave-length dial. AUTOMATIC VOLUME CONTROL. Walnut Cabinet.

£24 *Cash Price*

NOV. 11 PLEASE GIVE A LITTLE MORE REMEMBRANCE

MURPHY

MURPHY RADIO LIMITED, WELWYN GARDEN CITY, HERTS. TELEPHONE: WELWYN GARDEN 502

Printed by NEWNES & PEARSON PRINTING CO., LTD., Exmoor Street, Ladbroke Grove, W.10, and Published for the Proprietors by GEORGE NEWNES LTD., 8-11, Southampton Street, Strand, London, W.C.2, England.—November 10, 1933.

that in a cut-throat industry they could also trust one another. It was not unusual for one dealer to spare his own valuable staff and time to help out a fellow dealer stricken with illness at the crucial pre-Christmas rush period. These men felt that they belonged to an exclusive club or brotherhood, in which miraculously they were encouraged to think for themselves; to join Frank Murphy's crusade to clean up the radio industry; and to find life more strenuous but far more exhilarating in the process.

As might be expected, the dealers varied considerably in personality and background. Some already had thriving businesses in the centre of big cities; some were keen youngsters in shabby back-street premises who were determined to do better. Two, at least, were not strictly in the radio trade at all—one was a baker, and another an ironmonger. But they met Frank Murphy's requirements—they were honest, willing to learn, and determined to give good service to the public.

The Depression Years of the early thirties were at their grimmest just when Murphy Radio was struggling to get established. Competitors' sets were mostly available to the public at cut prices, through "friends in the trade", while Murphy prices were fixed and rigidly enforced and sets were available only from Murphy dealers and not from any unapproved radio shop. Matters became desperate when a certain manufacturer arranged for thousands of sets imported from the Continent to be sold in cheap bakelite cabinets at a price far below that of comparable British sets. So in November 1932, Frank Murphy summoned the representatives, each accompanied by one of his best dealers, to a confidential meeting in the "Cherry Tree" at Welwyn Garden City.

When all were assembled, he had the doors locked, and began to give them the stark facts. "If we are going to survive, we have got to cut our prices, cut our margins, cut dealers' discounts and sell a lot more sets to cover our own and dealers overheads. I've already approached the staff—they've all agreed to a 10% reduction in wages—that includes myself, by the way—and now I've asked you here to ask you to accept a lower discount than the normal $33\frac{1}{3}\%$—I suggest 25%. This will enable us to reduce the price of the *A3A* from £19 19s. 0d. to £12 10s. 0d., and the *B4* and new *A8* pro rata. Now, how do you think the sets will sell at these new prices?"

The first response was shocked silence. Then the various dealers came out with their known selling costs, and as these seemed to average between 20% and 24%, a discount of 25% was clearly not acceptable. The arguments went on for hours, with intervals for beer and sandwiches, but Frank Murphy was not going to let them go until the matter was resolved. The dealers' proposition of 30% was not acceptable to him, either. At last, Clifford Stephenson of Huddersfield suggested a compromise figure of "27½%", and this was agreed to by most of the dealers. Frank Murphy turned to Ted Power and asked what this would do to the new prices. After a few passes on his slide rule, Ted announced that the prices would come out at £13 10s. 0d. for the *A3A* table model, £12 10s. 0d. for the battery model *B4*, and £24 0s. 0d. for the new *A8* superhet. So at 10.30p.m. the doors were unlocked, and the famous meeting ended.

It was far from the end, though, for Frank Murphy. He called a series of dealers' meetings for the next fortnight, ranging from Scotland to Plymouth, to put across to the main body of dealers the same hard economic facts and the decisions taken to combat them. It says much for his personal charisma and the excellence of his products that "27½%" was indeed accepted as the dealers' working discount thence-forward. Clifford Stephenson recounts that when he got back to Yorkshire, he was accused of having 'sold the pass' on the question of discount, instead of being congratulated on achieving the extra 2½% which he had fought for.

Mrs. E. Trinder of Saltburn recalled the meeting with dealers from the Newcastle region which she and her late husband attended. Frank Murphy put it to them that though the discount on each set sold would be lower, it would be more than compensated for by the vast increase in sales. Some of the big dealers said they would like to discuss the matter without his presence. Mrs. Trinder said, "You could see that Frank Murphy did not like this, but he and his representatives left the room. The dealers then discussed the cuts and took a vote, resulting in a majority 'not in favour'. Frank Murphy was recalled and told the verdict. His reply was to request that all who had voted against his proposal should leave the room, as they were no longer Murphy dealers. This was very courageous stuff—or was it high-handed? It was the talking point in the trade for many a day. How thankful we were that we had not been swayed, and were still Murphy dealers!"

The 27½% discount remained a bone of contention for years, and in fact, was ultimately raised again to 30%, but by that time Murphy Radio were securely established as leading manufacturers in the radio industry and there was no question of them going under.

Those Murphy Dealers' Meetings—not just the one on 27½%—have become legendary in the radio industry, and for very good reasons. As Clifford Stephenson said, each year there was a new theme, a new idea to be thrown into the debating arena. Frank Murphy would start by outlining the Company's policy and programme of models for the coming year. Then he would try out his latest thoughts on the subject of retailing, or relate an encounter with a particular Murphy dealer which had warmed his heart—or roused his fury! Various technical experts such as C. R. Casson, Sir Gordon Russell, E. J. Power, E. W. Kent, P. K. O'Brien and Stan Willby would be invited to talk on their own special subjects. Then the questions would start. Back would come Frank Murphy's answer—sometimes slow and ruminative but generally a fast low-pitched delivery which left his opponent stumped for a reply, or an unexpected high lob which scattered those standing on the base line. A serious question was always treated seriously; but he certainly enjoyed a mischievous leg-pull if the questioner was at all pompous. Witness the incident recalled by George Denton at a London meeting, when a somewhat pedantic dealer enquired, "What would you say, Mr. Murphy, if we did so and so?" (details forgotten). Frank Murphy rose, paused, then replied: 'I'd say, 'Bugger the dealers'!" and sat down.

Mr. Wilde of Ilford once asked Frank Murphy how he would define "integrity", pointing out that the dictionary defined it as "uprightness, virtue, honesty, soundness". Frank Murphy agreed to the definitions given, but clearly rgarded it as a dynamic, not a static virtue. "I find you have to *learn* honesty, just as you learn any subject. I don't find, and don't expect to find, a manufacturer, or a dealer, or a member of the public of *perfect* integrity; but I expect to find, and indeed do find, those who are *learning* to be honest, and usually they are the first to admit that it is a slow and painful process."

Mr. Wilde then put in a searching supplementary: "Is it a

sound or honest policy to institute price-cuts and price-advances at only twelve hours' notice?"

'Well", said Frank Murphy, "Is anything sound or honest in the absolute sense of the word? As I see it, the fair question is, is it the soundest and most honest policy of which one is capable at the given time? . . . The arguing out of that particular policy took a good six months. No doubt, we ought to have been so much clearer-minded that we could have settled the question earlier, and then there would have been time to give both the dealers and the public much longer notice. But we accept ourselves as we are, and note for future reference how long what now seems to be a comparatively simple thing, actually took to work out . . . . Since, as I see life, no perfect decision is ever arrived at—and equally that decisions must be made—it follows that certain plans (say, for example, dealers' schemes) may be scrapped at a moment's notice, so that a bigger issue can emerge."

Other dealers, too, resented the lack of consultation on price changes. Mr. W. J. Haden of Swansea suggested that it would surely have been possible to pick twelve dealers from different parts of the country, and pledge them to secrecy; to which Frank Murphy replied shrewdly, "Shades of democracy! Suppose you, Mr. Haden, were not one of the twelve!"

Haden was supported by T. H. Colebourn of the Isle of Man, whose comment was: "I honestly don't think that it was a wise move to reduce the prices of current models. I accept fully Mr. Murphy's reasons".

He was rewarded with: "Oh, Mr. Colebourn, what a nice way of saying I am a fool, and that you'll be sporting enough to accept my foolishness—and do you really mean this?"

What Mr. Colebourn did mean, was that there were other ways of disposing of the increased profit; why not, for instance, distribute it among the staff and factory hands?

"They have not been forgotten", said Frank Murphy. "The average wage has gone up considerably since we started in 1930, and the trend is still upwards." (See Appendix, page 196). "When an economy is assured, it belongs to three groups of people—the consumers, those engaged in the work, and the share-holders. No one group is entitled to the lot, and the advantage announced on January 11th is the consumers' share."

Of course, the price reductions they were discussing in 1935 were not nearly as sensational as those which followed the famous "27½%" meeting two years before—then, according to Clifford Stephenson, Murphy Radio and their dealers made a most unusual gesture. They worked out a scheme of financial compensation for customers who had bought their sets at the original higher price, and refunded them in cash. No similar case was known to him.

The dealers' meetings were, therefore, the outstanding event in the Murphy Dealers' calendar. The idea was later taken up by other manufacturers, but on a simple promotional level, and since they lacked Frank Murphy's direct and democratic approach, their meetings were nothing like so provocative or enjoyable. In these days of "shuttle diplomacy", it is perhaps worth recording that in the early days, of Murphy Radio, Frank Murphy not only conducted and spoke at fourteen meetings in sixteen days, but also drove the considerable distances between the centres where the meetings were held. Rupert Casson was unwise enough to share a bedroom with him at one of the hotels they stayed in, and recalls being made to sit up till the early hours of the morning, discussing some fascinating points (to Frank Murphy) which dealers had made at that evening's meeting. Yet next day he was fresh and ready for the day's drive, and the evening meeting found him as full of ideas and enthusiasm as before. One front cover of a *Murphy News* has a full page photograph of the Man with the Pipe, with the caption underneath, "About these meetings—don't forget I'm listening to *you* this time!" That summed up his belief in the dealers' potential.

Many a dealer has subsequently confessed that he "never knew he had it in him" until Frank Murphy came along to offer him the challenge of the Murphy dealership; which entailed, as we have seen, far more than selling a particular brand of wireless set. Not only had he to spend considerable time after business hours in analysing the conduct of his own business and methods of improving it, but he was being driven to examine his innermost motives in his behaviour towards his customers, his staff, and the world in general. It was not enough to have banner headlines proclaiming "Murphy dealers are people you can trust"—every Murphy dealer was expected to live up to that claim, and like Nelson's sailors, they generally did so. Clifford Stephenson of Huddersfield, who in 1931 was one of the first to become the Sole

Murphy Dealer for his area and later successfully ran branches in Bradford and Halifax, was a keen member of the West Yorkshire Murphy Dealers' Association (which continued to meet all through the years of World War II); he remained in the industry until 1957, when he sold his businesses and went into local politics. He had a very enjoyable second career in local government, culminating in being made a Freeman of the Borough of Huddersfield. Today he freely acknowledges his debt to Frank Murphy and Murphy Radio. "I, like many another young man who came under Frank Murphy's influence in our early days, owe a debt to his teaching and inspiration. He made us think—before we were too old to learn and change. I salute him."

How it felt to be an appointed Murphy Dealer is well described by E. R. Worley of WORLEY & SEWTER of Gateshead. He wrote an article called "Our First Year" for the *Murphy News* (Sept. 8th 1934). It is worth quoting in full:—

"Having just completed our first year as Murphy dealers, the writer decided to sit down and try to find out what benefit, if any, our firm have derived from the dealership.

The answer, broadly speaking, is that it has done us untold good, but not, I should like to emphasise straight away, so much from the cash point of view as from the mental one; the cash return will come in full, I am convinced, when the aims of Murphy Radio are achieved.

I now firmly believe that, previous to dealing with Murphy Radio, our firm never knew what it was to think. We must have just carried on our business in the ordinary haphazard way which seems to apply to the majority of radio dealers. Such things as value for money, being servants of the public and so on certainly never entered our heads.

The first lesson derived from our dealership was that we should not concern ourselves so much with what cash profit we were going to make, but with the fact that we were offering the public the best value for their money, and that we would automatically receive a fair cash profit, providing we adapted ourselves to suit the conditions necessary to be a successful Murphy dealer. As Murphy Radio have said, no man can honestly maintain that it requires a 33⅓% discount to make handling radio a paying proposition. I am in perfect agreement.

Secondly, it was clearly shown that we must build up as quickly and soundly as possible, so that, as Murphy dealers, our buying public could look to us and be certain of receiving a square deal and sound advice.

Thirdly, there comes the personal touch which surely can only be found in Murphy Radio and their dealers. *A sort of brotherhood exists among the dealers*, at least certainly among those with whom I have come in contact. There is a spirit of mutual helpfulness. The same thing, of course, applies to the personnel of Murphy Radio.

Now, getting down to the commercial side of a dealership. Our firm have carried out moves and made alterations which I am quite sure we would not have done previously, for the simple reason that we only started to think on sound lines after obtaining the dealership; and the result is that our turnover, although possibly small compared with that of other dealers, is certainly moving along the right road. We now realise that to be a successful Murphy dealer we have a job of work on hand that is going to mean continual persistence towards 100% efficiency on our part, and thereby better value for money to the public (and incidentally a larger turnover for us).

Well, that is a summing-up of the result of our first Murphy year. We hope Murphy Radio are satisfied. We ourselves are, except that we hope to be more so, in another twelve months from now!"—E. R. WORLEY.

Many other dealers had reason to thank Murphy Radio for boosting their profits. In the editorial of the *Murphy News* of October 29th 1934, we are told:—

"It is worth recording that *in the last twelve months over a hundred dealers have taken new shops or opened branches, many of them in the best shopping centres*; an equal number have carried out structural alterations and improvements to their existing premises, while roughly another fifty have added substantially to their equipment for the better service of the public."

A typical example was that of H. YORK & SONS of Kettering, who had occupied a small single-fronted shop from 1922 to 1930. They then moved to a larger shop, and were appointed Murphy dealers in 1931. When this shop proved to be inadequate, they bought a site which included a disused factory, and built a much larger two-storeyed shop on this prominent corner site in 1934. Pictures of both old and new shops appeared in the *Murphy News* soon afterwards.

Another case is that of T. A. ROWNEY & SON of Stafford. Tom Rowney began by enthusiastically selling Murphy sets as the sales manager of a Murphy dealer who did not stay the course. Determined to secure the dealership for himself, Rowney persuaded a fellow employee, Bill Pearson, to put up £200 to match his own slender resources. Between them they bought some of the former dealer's stock, acquired a vacant shop from the local Council, and formed a new business. It never looked back. Today, in his 'seventies, he is still active as Managing Director of two flourishing shops, with his wife as Company Accountant and his son Peter (now a Director) ready to take over on Tom's retirement and carry on the old Murphy tradition of service to the public.

ALEC VALLANCE, another of the early Murphy dealers, having worked in Mansfield for some years, decided to break new ground with the promise of the sole Murphy dealership by opening a radio shop in the main street of Scunthorpe. Illustrations in the *Murphy News* show the shop front before and after alterations. In the second picture a large blue fascia has in huge letters the message *YOUR MURPHY DEALER* while in the window below, Murphy sets are boldly displayed. (This was only the first of many original and eye-catching displays). Alec Vallance ensured that the 60,000 people in Scunthorpe and the sur-rounding district were made fully aware of who and what he was, by taking large advertisements in the local paper every week. As a result, he sold £1400 worth of Murphy sets in his first three months; and this was just the start of a highly successful and 100% "Murphy Mad" career.

Another well-documented history concerns CUSSINS & LIGHT of York. In 1934, Reggie Cussins and Pat Light were trading at 34, Walmgate, some distance from the centre of York. Frank Murphy then announced that in future only one dealer in York would be retained as the Murphy dealer. There were other bigger retailers in the city, but he thought that Cussins & Light could do the job best. He offered them the dealership, providing they moved to a central position in York. Taking a calculated gamble, they moved to No. 1, King's Square (which Frank Murphy visited on the opening day), and immediately experienced a dramatic increase in business. As their new premises were large enough to accommodate a service department, Cussins and Light asked Murphy Radio for advice on its layout. Not only did the Company offer advice, they sent George Berry from Murphy's own technical staff to organise it personally. Cussins and Light were so impressed with his work that they offered George Berry the post of Service Manager with them. He talked it over with Frank Murphy, who, keen as ever to forge links within the organisation, encouraged him to accept. So for six years George Berry stayed in York, until wartime direction of labour sent him back in 1940 to Welwyn Garden City, taking with him the new Mrs. Berry, who was Pat Light's cousin. He remained with Murphy Radio until he retired, having succeeded D. W. Parratt and R. O. Seccombe as Head of the Service Department there.

Meanwhile W. D. (Denys) Cussins, son of the co-founder of the firm, had left the local secondary school and graduated from Cambridge at the age of nineteen, taking a Double First and coming out top of his Tripos. After some years in academic work, he returned home in 1953 to join the family firm. In 1971 he commemorated the Company's Silver Jubilee by writing its history—a unique account, not only of the Cussins & Light group of companies, but of the development of York itself through those years, and of the pioneers of electrical appliances and communications systems, from Edison and Marconi onwards.

It is therefore all the more heart-warming to find Denys Cussins in his second chapter paying special tribute to the subject of this book:—

"The one single event which sent the business rocketing upwards was the appearance in the radio industry of Frank Murphy.

In September 1929 Frank Murphy, aided by a staff of seven, began producing radio sets in a small factory in Welwyn Garden City. The sets themselves were very good, but that was not what distinguished the Murphy Radio company from many others in existence at the time. The great difference was the driving genius of Frank Murphy. The products of his factory were only the start of the deal he offered retailers.

He believed in fixed retail prices at a time of special offers, etc.; (yes, just like the situation today). On the other hand, he believed in reasonable, not excessive profits. As a result, the discounts he offered were fixed, irrespective of quantity, and lower than any others in the industry. Furthermore, he insisted upon the highest standards of service, in all senses of the word, from his dealers. It was a system which was eminently simple and just to all concerned, including the public."

In Denys Cussins' opinion, his Company's decision prompted by Frank Murphy, to move to the centre of York, was probably the most important they ever made. By 1971 they had expanded enormously, with five satellite companies, and retail outlets in twenty towns throughout the surrounding area of York, and had achieved a national reputation in the radio-electrical industry.

As for their "Murphy Madness", they were willing to go to almost any lengths. When Murphy Radio brought out their giant Frank Murphy heads for carnivals, Reggie Cussins paraded through the streets of York wearing one of them accompanied by Pat Light dressed as a "dotty Professor", offering a prize for the best photograph of them by a member of the public!

More seriously, they gave practical expression to the Murphy philosophy by inaugurating in 1944 a profit-sharing scheme for all members of staff, whereby 3¾% of the yearly profit on capital was distributed among them, with appropriate weighting for length of service, while a similar amount was ploughed back into the Company.

As one reads through the first editions of the *Murphy News*, the early Murphy dealers spring to vigorous life. What made them—the best of the nine hundred, at least—throw themselves so whole-heartedly behind Frank Murphy, with his products and his philosophy? As another veteran dealer, Frank Thompson of Sunderland put it, "Frank Murphy was far in advance of his time. Only now—over fifty years after he founded Murphy Radio—are people beginning to give general credence to many of his ideas. We couldn't follow all his theories. But we knew he was a man to trust—and he made us feel that life ws worth living, however hard the going".

But there was another side to the story. Some people who knew Frank Murphy well were beginning to have misgivings. True, the Company and all associated with it were enjoying an unparalleled success both financially and in terms of prestige, but the trouble with people who have exciting ideas is that they are constantly dreaming up new ones, and Frank Murphy was no exception to this rule.

Satisfied that the brilliant team at Murphy Radio could be trusted to carry on the main tasks of designing, manufacturing and servicing Murphy sets, Frank Murphy turned his attention in 1934 to the one aspect of the industry which had not yet been subjected to scientific appraisal—its retail distribution system. It was clear that a few outstanding dealers were doing far more business than the average, and he wanted to know why. Was it their own personality and drive, or was it a purely fortuitous circumstance, such as the siting of their premises, which resulted in so much higher turnover? Was a central location in the main town shopping area necessary, or could an off-centre site be equally successful if the dealer redoubled his publicity so that people knew where to find him?

In an effort to find a rational answer to this question, Frank Murphy spent many hours examining and comparing dealer's turnover figures in relation to their outgoings, and in 1934 he first put forward his famous Rent Theory (later renamed the Turnover Equation), in which he established the vital relationship between the amount of rent paid by the dealer for his shop, and his subsequent turnover. The formula he eventually produced was as follows:—

"TURNOVER=27.2×RENT+13.7×ADVERTISING+8×CANVASSING"

Naturally, he lost no time in putting this new idea forward at the next series of dealers' meetings, pointing out to them the enormous advantages they would gain by moving to the most central (and therefore most expensive) High Street shopping site—which he called the "Woolworth site". Such were his powers of persuasion that over a hundred dealers took his advice and made the move, with a very rapid increase in turnover, as he had predicted.

Another aspect of retailing was the personal element. Did the customer need or welcome advice before making his purchase, or was the relationship with the retailer limited to the handover of the goods in return for cash? Should the transaction end there, or did the customer need a guarantee of replacement of defective goods and the promise of after-sales service? Was the retailer acting on his own behalf, or that of the customer, or was he primarily an agent of the manufacturer? If there was a relationship between the three parties, how could it be defined?

Probing ever more deeply, Frank Murphy asked himself, "What makes the customer walk into the shop in the first place, and once in, what makes him either buy, or walk out again?" And so on, until he was down to the fundamental question, "What is a shop?" He took a mischievous delight in firing this innocent-seeming enquiry at friends and colleagues. If they replied, "A shop is a place where you buy things", he would counter with, "Is a railway booking office a shop, then?—Or an ice-cream van?" He felt that if he could first establish the basic function of a shop, a coherent theory of distribution could be developed.

By the end of 1934 Frank Murphy's unusual views on advertising and distribution were becoming widely known, and he was invited in January 1936 to address the Publicity Club of London, and to contribute an article on "What is a Retailer?" (published in *The Broadcaster* on March 28th, 1936), as well as to take part in a BBC series of talks and discussions on retailing.

In all of them he began by defining the word "shop", which he said could cover three quite different types of trading:

(1) a manufacturer's agent or branch
(2) a minder of a dump
(3) the consumer's agent

all three being humorously illustrated in the article in *The Broadcaster*.

Frank Murphy pointed out that the first group—those shopkeepers who were operating as manufacturers' agents—were largely dependent on the scale of the manufacturers' advertising, and the greater that was, the more their businesses were likely to thrive. By the nature of things, they had to be biassed in favour of the product they sold; they could not give the customer independent advice which might lead to his decision to purchase a rival product. They *could* be honest, but it was more likely that they could be making money by exploiting the customer's ignorance.

The second group of shopkeepers, he said, were those tens of thousands of small traders who in essence were no more than minders of dumps of goods, which they rather pathetically hoped to be relieved of by kind-hearted consumers. Such people used advertising—if they used it at all—in a vague, purposeless way. While not deliberately exploiting the consumer, they certainly could not be said to serve him in serious intention.

The third group of retailers, however, were different. They could be described as purchasing agents for the consumer, if they filled the following conditions by offering:

> (a) *a choice of price levels and specifications* (e.g. for the customer who asked to see "a radio set costing about so-and-so").
>
> (b) *Qualified staff* who could assess from personal knowledge the comparative merits of different makes
>
> (c) *Integrity on the part of the staff:* to give the consumer honest advice and so protect him against exploitation.
>
> (d) *Convenience of locations*—important to the consumer for his simplest everyday needs, but of far less importance than (a), (b) or (c) in the case of a costly durable such as a radio set.

This was all very fine, but during 1935 the rumour went round that Frank Murphy was considering the relative advantages and disadvantages for Murphy Radio, its dealers, and the public, if Murphy dealers became, in effect, Murphy branches. Strongly impressed by the good reputation of the Ford Main Dealers, and noting that they had set up a branch in London's Regent Street, Frank Murphy charged George Denton of his distribution department with the task of investigating whether Murphy Radio could not acquire some similar shops and so deal direct with the public. Although the idea came to nothing, the mere rumour of it sent cold shivers of apprehension down many dealers' backs. At the same time, Murphy Radio were to experience some serious competition from another new company, Bush Radio, who, under their very able Sales Director, Darnley Smith, were offering a Limited Dealership similar to Murphy's but with a larger discount. As a result, a number of the bigger dealers decided to hedge their bets. They took up the Bush dealership, and naturally, their sales of Murphy sets dropped. The Bush sets were good and reliable, they were housed in pleasant, if undistinguished cabinets, and they sold well. It was not a good time for Frank Murphy to be shaking the hitherto unbounded confidence which his dealers had placed in him and his company.

The snag for a researcher who is also a brilliant expositor is that whatever theory is currently being put forward, the audience naturally assume that the researchers believes it to be true—regardless of the fact that, a week or two later, a totally opposite theory may be argued equally convincingly. Such rapid changes of tack came naturally to Frank Murphy—so much so, that his own family nicknamed him "Toad of Toad Hall"—but this did not affect his ultimate objective, to get at the truth by every means possible. To those who did not know him, the fact that Frank Murphy was circulating a discussion paper on "Branches or Dealers" meant that he had already made up his mind on the matter, despite his estimate that it would take twelve months' hard arguing "before we see daylight—and even then the job will only have been started".

One of the dealers who was obviously quite convinced that Frank Murphy was aiming at branches was H. Knowlson of Abergale. He sent an open letter, saying:—

"Many Murphy dealers are now seriously alarmed, and I venture to say that 75% of them would sooner struggle along, eking out an existence on their own, than make another few pounds per week working for someone else. Now I am going to be very blunt—I suggest you have published your '27 thoughts' (or queries), with the request that dealer should go into the matter fully at the next meetings, because you know that you will not have the slightest difficulty in answering and flattening any opposition that may arise; you could do it in such a way that you would no doubt feel justified in carrying on with your scheme. *A man like you has made up his mind what he wants long ago.* But to float successfully a new company (say, Murphy Distribution Ltd.) and get hold of so many businesses for so much share scrip you must persuade dealers that the present method of distribution is wrong, and that the inevitable end must be Murphy branches. If you get away with it, you will be the smartest man in the radio trade!

I have had the pleasure of speaking to you on two occasions, and have listened to you at a number of dealers' meetings. My impression of you is that you ooze personality and are quite capable of leading a large number of Murphy dealers up any "Garden City" path, unless in the meantime they come out of the trance into which you have put them.

Now, I suppose, I shall get a 'kick in the pants'!"
H. Knowlson. (SLATER & WHEELER).

Alec Vallance, on the other hand, after a thoughtful letter discussing Frank Murphy's arguments, finishes on a more personal note:—

"Dealers are asked to consider 'What is your aim in life?' Well, I have altered the question a bit, and I ask, 'What is my job?' The answer I give myself is, 'My job is to help to make the consumer's pound go as far as possible'. It is a simple answer and easy to apply to problems that arise in shopkeeping. If I am sincere in giving this answer, then any system of distribution which would enable me to give more for what I take out would be welcomed by me.

Personally, I would be pleased to take my chance in Murphy distribution. I would very much dislike serving customers in a Murphy shop, but I dislike doing that in my own shop, so that would be all square. Besides, one is naturally optimistic!
Alec F. Vallance. Scunthorpe."

But undoubtedly the letter which must have pleased Frank Murphy the most came from James Nichol of Edinburgh:

"I recently spent many hours in an attempt to express my views on 'Dealers versus Branches'. After reading over what I had written, I decided to scrap the lot in favour of a brief summary. In my opinion there is only one answer to the whole vital question; *Dealers must assume the position of purchasing agents for the public, and remain totally independent of any one manufacturer.*

I think we are all agreed that success or failure depends entirely on our attitude towards the consumer; and if we employ integrity, in every sense of its meaning, by honest unbiassed advice, we are serving the consumer as he desires to be served, thereby assuring for ourselves our measure of success.

If, therefore, Murphy Radio decide that Branches will be their new road, then MAITLAND RADIO will be parting at the crossing. If you think we have become damnably honest, then blame Murphy Radio Ltd."

Actually, Murphy Radio decided that Branches were *not* their new road, and the dealers breathed again. But Frank Murphy's capacity for putting deeply provocative questions to them was by no means exhausted, as we shall see in the next chapter.

A generation later, in 1955, Geoff Dobson of Huddersfield looked back to what his contemporaries termed "the Bad Old Days":—

"... Certainly we all worked ridiculously long hours for very small tangible reward. But surely, life was good in those days. There was a sense of achievement in the rapid expansion of the industry and in the quick technical developments which were being mastered. How pleased we all were to master the technicalities of mains superhets, automatic volume control, automatic tuning correction, and the multitude of new ideas for obtaining the last word in high quality reproduction! In our shops, clever display ideas were evolved, made and put into use, Service departments were fitted out with impressive pieces of test equipment, much of it made by our own good hands.

Yes, our wives and sweethearts gave us many a dressing down for staying 'at the shop' into the small hours, and accused us of being more devoted to *The Wireless World* and its contemporaries than we were to them, bless their hearts. In those pre-war days we had our problems and spent a lot of our spare time thinking about them and, yes, in the main the answers that we found seemed reasonable—that is to say, they were *considered* answers which did not cause us any misgivings—and most of them have stood the test of time. *But, more important, during our working life we seemed to have time, when a problem raised it head, to pause and have 'A bit of a think' about it."*

Bad old days? Not a bit of it! The long working week, month in, month out, year after year, still left us our sense of humour and a sense of something achieved. More important, though, we still had time and inclination for a natter with our contemporaries about our job!"

This ability to "take time off to think about it", Geoff Dobson suggested, stood those dealers in good stead who were left to struggle through the war years short of staff and resources, and it also enabled them to face the post-war challenges of television and the ever-increasing mound of paper-work. He reminded his fellow dealers that both they and their businesses would be revitalised if they made sure that everyone, at whatever level, had "time and inclination for thought".

"Exchanging thoughts about one's job" was indeed one of the pursuits of the typical Murphy dealer, and it was soon evident that the keener ones wanted to meet together on a regular basis. So the first *Murphy Dealers' Associations* sprang up, to become a unique feature of the radio industry, and one that was to foster friendships that deepened over the years. The Newcastle dealers formed the first Association, and they were soon followed by dealers in the North and West of Scotland, in West Yorkshire, the Midlands, the North East, the North West, Northern Ireland, the Eastarn Counties, Greater London, the South East and Wessex. With members launching forth on topics of mutual interest, there were many hot debates on the principles and practice of retailing, but always in an atmosphere of friendly good humour. "Ladies' Nights" followed, and some of the ladies developed a good line of their own—at least three were dealers themselves, and many wives, like Mrs. Rowney and Mrs. Trinder, were thoroughly involved in the businesses, so they, too, had a point of view to express. In this way the various Dealers' Associations helped to perpetuate the Murphy tradition of integrity, service and technical innovation.

As the founder members of the first Murphy Dealers' Association grew older, there was a general feeling that they must hand on the Murphy tradition to the younger people growing up in the industry. As Jack Hum, editor of the *Murphy News* from 1955, put it: "Those who were young when Murphyism was young have a special duty to ensure that they are not the first and last of a line. There should be others following along after them who can show that the future is in good hands. If there aren't any such in a given dealership, then something ought to be done about it."

The "something", he suggested, should be the formation of junior branches of the Murphy Dealers' Associations, giving the staff of Murphy dealers their own organisations, rather like the Round Table Movement which grew out of the Rotary Clubs. In fact, West Yorkshire had already in 1955 paved the way with its own Senior Staff Group, and Geoff Dobson, whose views have already been quoted, was a founder member of it.

Now that Murphy Radio no longer exists, there cannot logically be any Murphy dealers. But the old spirit still persists, and the stories of the "good old days" continue to be told, as old friends meet. As for the affection and regard in which Frank Murphy was held by his dealers, I still have the gold watch which the Newcastle Dealers' Association presented to him, inscribed "To an Idealist—from Northern Friends. March 8th, 1937."—with a crude but recognisable portrait of a man lighting his pipe!

Chapter Eight

"SPREADING THE GOSPEL"—the *Murphy News*

"Give me the liberty to know, to utter, and to argue freely according to conscience, above all liberties." *John Milton:* Areopagitica.

Three years after launching Murphy Radio, Frank Murphy came to realise that it ws physically impossible for one man, even with the help of a score of representatives, to maintain close and regular contact with the thousand or more Murphy dealers and the six hundred people in the work-force at Welwyn Garden City. The conventional radio manufacturer, who needed only to announce, "Here are our new models for this year", could be content to send an occasional circular to his dealers, but in Frank Murphy's view, Murphy dealers needed to know not only what they were selling, but also how, and even why.

With a few shining exceptions, they had to be taught how to make the most of their staff, their premises, their local contacts, and their service departments (if they had any). They needed to be encouraged to think about their work, and to organise their businesses as efficiently as possible, having as their principal aim "service to the customer", not the highest possible profit for the least effort—though profits would certainly come to those who were prepared to work hard and intelligently. Above all, the dealers had to be infected with his own fierce enthusiasm, to "get it in their blood", as he expressed it later; and while he could do this admirably at the famous Dealers' Meetings of the 'thirties, he could hold these meetings only once or twice a year. Moreover, some of the dealers, for various reasons, might not be able to attend and get the "Murphy Madness" from its source. Some other device needed to be found to carry the message and to relay the responses.

Thus, on May 29th, 1933, the first edition of the *Murphy News* appeared, to be published fortnightly and mailed free to every dealer. Internally, it was distributed to every member of the workforce. It was to prove a winner from the start, largely because, as always, Frank Murphy had picked the right man to be the editor. Typically, he went for the top man in the field—Stanley Willby, then editor of *The Wireless Trader*; and equally typically, having given him the job, he left him to get on with it without interference.

Not without warning, however. The first page of the new house journal contained the following succinct message:—

"We believe that every one of our activities should stand or fall according to whether it ultimately serves the public or not. The *Murphy News* will be no exception to this rule. If it succeeds in binding Murphy dealers and Murphy Radio closer together in a more effective appreciation of our common problem of becoming better servants of the public, then the *Murphy News* will be allowed to continue. If it does not, but is merely a house organ which it tickles our vanity to have, the sooner it is stopped, the better.

FRANK MURPHY

The first article in the *Murphy News* was devoted to a subject of vital interest to all Murphy dealers. Entitled "THIS SUMMER SALES BUSINESS", it opened cheerfully enough with the news that in its first three months of production over 6,000 of the new model, the *A4* superhet had been sold, while sales of the D.C. variant and battery models were encouraging, but could be better. This led to the main message to the dealers—warning against being complacent. The article pointed out that though Murphy dealers had sold 3,000 sets in March, other (non-Murphy) dealers had sold 17,000 non-Murphy sets. Consequently, Frank Murphy had sent a letter to all his representatives telling them that *"Dealers must be prepared to go 'Murphy Mad', or go on the stop list . . . There is no reason why we should spend £70,000 a year on advertising and put up with lukewarm dealers."*

The article on summer sales went on: "It must be taken as a fundamental, though possibly uncomfortable truth, that when an organisation grows, the people in it must grow with it. There is no alternative between that . . . and being left behind".

The point was reinforced by the bald statement that in one important Midland industrial town *all* seven dealers had been put on the stop list (i.e. sacked) because they were not pulling their weight, though some might qualify to be reinstated later.

Just in case the dealers who read the article did not apply the moral to their own case, they were advised to study a table on the back page, where monthly turnover figures were set out for each area in the country, and showing for each area both the *average* dealer's turnover, and the highest *individual* dealer's turnover—usually three or four times the average.

Other important contributions to this first *Murphy News* included a technical explanation why the Company had chosen to use "Positive Drive Class B" output in the battery sets, using Mazda triode valves; a page for the Service Man; a page on the possibilities of local advertising by dealers (copy supplied by Murphy-Casson Ltd.); and various illustrated articles, all with a practical bent, on window displays, showrooms, giant heads for sandwich-men and other publicity gimmicks, with numerous titbits of information and ideas from fellow dealers. A much lighter note was struck by reminiscences of the first days of Murphy Radio from Sidney Carne (the first Works Manager) with cartoon-like illustrations, while the works employees were entertained by a picture of the victorious 1st cricket XI of Murphy Radio and an account of their matches.

Typical of the magazine's amnusing but hard-hitting approach was the following small "fill-up":—

R.I.P.
J.D. Drysdale,
Murphy 'rep' in
Kent, Surrey and
Sussex, recently felt
constrained to cancel a
South Coast dealer's
appointment. His report
to Sales Office was
brief and to the point.

Reasons for cancellation
No enthusiasm.
No stock.
No brains.

No hope.
To which we need add
only one phrase—
Verb Sap!

One can imagine the excitement and trepidation in the little editorial office as reactions were awaited on the new publication. These were speedily forthcoming; in the next issue a selection of letters appeared entitled "Bouquets and Brickbats". One from the dealer in Rossendale was typical:—

"I wish to express my thanks and appreciation for the first number of the *Murphy News*. It is a step in the right direction—that is, the binding together of directors, work people and dealers of Murphy Radio. The *Murphy News* will, I hope, be like the 'family circle', where we can all meet together to discuss the doings of the day, and outline future endeavour, and be sure of a sympathetic hearing. It makes one feel that one cannot work too hard as a member of such a family. In my opinion, the first number is a big success, and if I know Murphy Radio, the future numbers will be even better.

ARTHUR COULTON."

Several dealers wrote in to ask for two, three, even four copies of the subsequent issues, for the benefit of their staffs.

The format of Number 2 was similar to that of Number 1. Stan Willby's Editorial set the tone for an even stronger, livelier issue:

"*ON BEING RUDE . . . And Some Other Subjects too*

It is often said that we in Murphy Radio are inclined to be pugnacious. 'Domineering', 'belligerent', and similar terms are occasionally used to indicate the same opinion. Maybe it is because we have the habit of saying what we mean . . . and meaning what we say. In any event, no matter which way you look at it, it is interesting and instructive to observe the *effect* in certain instances of that directness which many people call plain rudeness.

"For example, among the territories covered by Frank Murphy in his recent chase round the Kingdom was (of course) Lancashire. Here are two extracts from the representative's report, with only the dealers' names omitted. The underlined parts are significant.'

Case 1: I sent back the order for two *A4*'s and told them I still awaited the suggested letter setting out a future policy. This came, and I went to see them. They acknowledged *having played about with wireless business* in the past, but said they had come to their senses and would, if permitted, put their backs into the Murphy dealership. They offered a good window-display, good stocks, and *an immediate order for eight sets.* I accepted this, and am seeing them again for further discussion.

Case 2. This dealer phoned me on the Monday after our Saturday call and said it had been the first time in his business career that a manufacturer had been in his shop and *told him something worth listening to.* He was going to do something about it right away. Watch!

He telephoned again on the following Friday, and said that he had that minute sold his fifth Murphy that week. He wanted me to *thank you for geeing him up* and would like you to know that his assistants were very bucked . . . He and the staff had personally *gone out, almost house to house to earn back their stripes*—and he would have to spend the day in bed on Sunday . . .

He is now having a new £300 shop front, and I have given him a branch dealership . . .

> "An ounce of help, we are always told, is worth a ton of sympathy—and that holds true even if the 'help' *is* in the form of straight and therefore possibly unpleasant talking.
> You may be interested to read the final sentence of that rep's report:—"
>> 'Thank you for being rude to the few dealers we managed to see; *they are all doing much better."*
> "Note that last part—*they* are doing much better. Is one to assume that *anyone* would prefer *not* to do 'much better' rather than to hear some occasional plain speaking?"

This *Murphy News* article went on to talk about:—

"**THE SALES CURVE:** All the dealers know (and some of the staff, unfortunately), that sales were down last week—an over-all drop of something near 20%, although the *B5* figures were up. This was the first time that the curve had shown anything but an upward trend since early March—and we feel that dealers ought to do something about it.

Let's first answer the inevitable question, though, namely—'What are *we* doing about it?' We are maintaining national advertising at a summer pressure never even attempted by any other manufacturer before, we are providing new sets that are outstanding value for money, and finally we are appointing twelve new representatives (making twenty-nine in all) in order to give dealers better service and more frequent attention and help even than in the past. Now, we may ask again 'What are *you*—and most particularly the Dealers—going to do about it?' . . ."

After a page listing the names and addresses of the new representatives, the next article described the formation of a unique body, *"Murphy Radio Circle No. 1"*, a group of Murphy users, founded in January 1932 by a few enthusiasts in Glasgow, who worked out their own rules and constitution, and met regularly for lectures, outings, whist drives and musical evenings. The founder and secretary, Mr. W. Simpson, reported that they now had an established membership of 75.

In other ways, the format of *Murphy News* No. 2 was similar to No. 1. There was a page for the Service Man, two more tales of the "early days" by representatives G. C. Proctor (Eastern Counties) and Hugo Herod (Northern Ireland), an illustrated article on shop signs, banners and other publicity aids, a dealer's imitation newspaper, and more examples of good window displays. The Works side was featured in a breezy description, with photographs of the Sales and Accounts Departments' removal to roomier offices at Bridge Road; and there was news of the forthcoming Annual Sports Day and the Annual Works Outing (to Skegness); a report of a cricket match between Murphy Radio and Murphy-Casson; and a faintly surprised but nevertheless warm tribute to the wife of the Market Drayton dealer who had managed to keep their business going smoothly when he was laid low with an operation. No Women's Lib then!

The most significant innovation, however, was the Correspondence Page. Here letters from six dealers on diverse subjects foreshadowed the future function of the *News*. Each of the writers had something positive to say: "Will Murphy Radio make car radios?" (Alec Vallance, Mansfield); "Should we start a Murphy Dealers' Association?" (L. Rosen, Croydon); "We can't

Fig 23. Frank Murphy with two of his dealers - F. Piggott of Kingston-on-Thames, and James Thornes of Dewsbury

Fig 24. Dealers Publicity (1)

Fig 25. Dealers Publicity (2)

MURPHY NEWS

IT'S CARNIVAL TIME

Mr. Thornes, of Dewsbury, spreads the Murphy gospel at Mirfield, a neighbouring township, with this decorated trailer. Below, left: Mr. Hoddinott (Wallingford) and Mr. Platt (Marlow) combine in an aquatic venture (story overleaf). Right: Winn & Hay's van shouted Murphy at Upminster Carnival. Bottom left: Scurrah, of Bradford, seeing the "giant A24" idea described in the "News" put it into practice.

Above: Mr. Vivian Morris, of Sutton, Surrey, (right) and Rep. Brothers, with a Bighead mounted on the latter's car.

Fig 26.   Dealers Publicity (3)

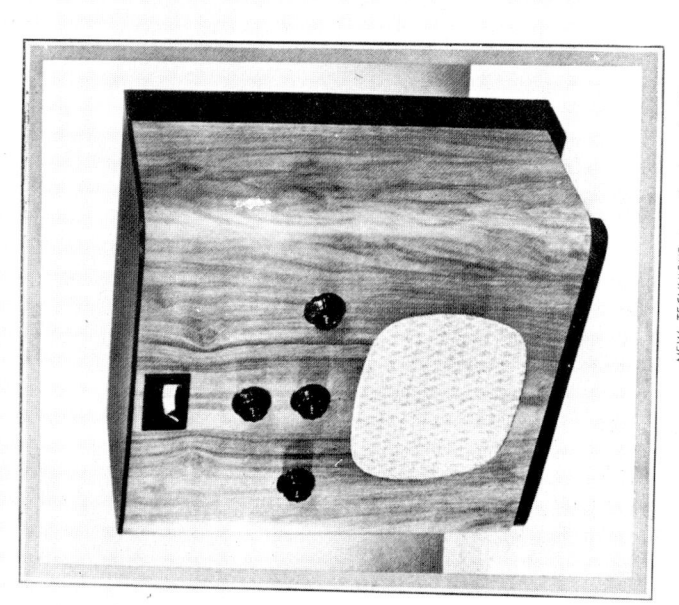

NEW TECHNIQUE is employed in the B27 cabinet, in the form of a single sheet of bent plywood for the front and sides. Another departure from former practice is the transposition of the chassis and speaker compartments in order to accommodate the weighty batteries at the base.

Fig 27. New Technique in the B27 cabinet

A CHINESE OVAL loud-speaker opening distinguishes the new cabinet design for the A30 and D30 table-sets. Walnut front panel, with a band of vertical-grain Zebrano across the base, add to the appeal. The sides and top are uniformly treated with a rich, dark-brown stain.

Fig 28. The 'Chinese Oval' of the A30 loudspeaker

Fig 29. The A34 (1937)

Fig 30. The B31 (1937)

lend you our dummy *A4* shown in the last issue, but we'll describe how to make your own" (C. Stephenson, Huddersfield); "We've scotched a customer's fear that Murphy's were going broke, by showing them the *Murphy News*" (Hills Bros, Upton Park, London); and two fairly trenchant criticisms, one from R. E. Manton of Clapham, having a swing at Frank Murphy's "Murphy's Mad" letter, and a more serious attack by I. Drazin of Hampstead on the topics chosen for current Murphy national advertisements.

Astutely, the editor did not give an official reply, but invited comments from dealers and staff on the subject, thus ensuring plenty of interest in issue number 3!

The front covers of the *Murphy News* were as striking as its contents. The cover of Vol. 1, No. 1, rather, consisted of a full page photograph of Frank Murphy, looking rather serious as he concentrated on filling his pipe. No. 2 had a line drawing of a newsboy with a dealer's postcard blown up to placard size, which logged six sales of Murphy sets on one day, plus the caption, "What can be done!" No. 3 showed a typical load of Murphy sets in their cartons being loaded for despatch to dealers on the platform of Welwyn Garden City railway station. No. 4 featured a Manchester dealer's brilliantly coloured Murphy dealer flag, with the caption, "High Flying!" . . . and so on. The year closed with Frank Murphy again (his face framed by an hour-glass) with the message "In 1933 the man-in-the-street has supported Murphy dealers and Murphy Radio well. Let's say 'thank you' by trying to give him an even better deal in 1934."

It was soon clear that Stan Willby and Jack Hum (who had by now left *The Wireless Trader* had hit on exactly the right mixture to capture the continuing interest of their readers. Technical articles, well-illustrated and written in simple language, were interspersed with write-ups and bouquets to energetic and imaginative dealers whose efforts might inspire the rest; forthright discussions on matters of policy, such as "100% Murphy or not", the whys and wherefores of not exhibiting at Radiolympia, justification for raising or lowering dealers' discounts, and the style and content of the firm's national advertising campaigns; and humorous anecdotes and human stories to lighten the long hours of making and selling the sets. The *Murphy News* had become the vivid expression of an extraordinarily live organ-

MURPHY NEWS

MARCH 24, 1934

VOL · 2 · No. 6

ONE DEALER PER SHOPPING CENTRE

isation, and as such was eagerly awaited and devoured by all its recipients.

What is fascinating about the *News* is the way in which its original aims were steadily broadened, and particularly how it developed as an open forum for the views of all in the Murphy organisation—which set it completely out of the general run of house magazines. Thus in the issue of January 11th 1936, seven whole pages in the centre were devoted to dealers' letters voicing strong criticisms of Murphy Radio's end-of-year price reductions for existing models, while on April 18th in the same year, six pages were occupied with staff comments on the recently announced holiday and sick-pay benefits (which were well in advance of current industrial practice), two pages were given to reprinting replies from non-Murphy dealers to Frank Murphy's provocative article in *The Broadcaster* trade journal on "What is a Retailer?", and two pages at the end of the issue gave a dealer's detailed criticisms of the cabinet design of the *A24* and *A26* models, and also designer Gordon Russell's answer.

The editorial of May 16th 1936—the third anniversary of the founding of the *News*—contained a characteristic stock-taking of its progress:—

> "If deaers are to be believed, most of them now regard the *Murphy News* as a permanent part of their landscape, and get decidedly grouchy if it isn't delivered until Monday morning. All sorts of outsiders who get the *News* for various reasons—from the titled nobility to a Sheffield tram-driver, occasionally pass favourable judgment in embarrassingly flattering terms. From the staff, though most of the criticisms which reach us are both pointed and destructive (as befits one member of a family to another), we still get the general impression that privately they are rather proud of the firm's journal, and find it more interesting than they are prepared to admit.
> So much for the credit side, but there is another side to the balance sheet... So far as we can gather, most dealers read these pages deeply enough to absorb more and more of this Company's outlook on life; but some of them don't watch for (or maybe it's just that they don't register in their minds) the routine bits of information that appear in every issue...
> As to the staff, no one is more conscious than ourselves that

their interests might have received greater attention and more space in the past. Recent months have seen a good deal of improvement in this direction, but it looks now as though further progress will largely depend on the staff themselves.

... in simple terms, we think *The News* is pretty fair, but could be still better. Any effort that will help us in the future to make the progress of the past three years look like a funeral procession will be very much appreciated."

Later in the same issue Stan Willby attempted to define 'freedom of speech' in the context of *The News*, emphasising that this was a first try, and urging others to make their own definitions:—

"Freedom of speech within Murphy Radio means that any employee is at liberty:—

(1) To comment on any aspect of the Company's activities if he feels he can offer constructive criticism.

(2) To discuss conditions within the Company which he has valid reason to believe are unsatisfactory to a sufficiently large number that they may have an adverse effect on the whole community (bad lighting, e.g.).

(3) To air a personal matter—even a grievance if desired—which he assesses involves a principle likely to affect others. (e.g. a case of alleged personal injustice, over which it has proved impossible to get satisfaction in the normal manner)."

While Willby did not get any "takers" in his challenge to others to provide an alternative definition, he did receive and publish a letter from Frank Murphy himself which showed that the Managing Director was fully aware of the dangers of "speaking out of turn" in the usual industrial organisation:—

"Some time ago I said to Willby, 'I should feel very happy if I could see member of the staff writing in the *Murphy News* as frankly and as seriously as the dealers write. He suggested in reply that if I waited a little while, I should see that happening—and now I'm seeing it.'

But it has come to me several times lately that some of you people are saying things to this effect. 'This idea of writing letters of the *Murphy News* is all very fine and large, but there's one thing which will always prevent it working successfully: namely, that anyone who does get it "off his chest" in the *Murphy News* will only find that his office-boss or his charge-hand has saved it up for him until the end of the year—and then out he goes!'

On reflection, I felt that this was very reasonable sort of gossip, whether there was real justification for it or not. To me it seemed also quite on the cards that the 'office-bosses and charge-hands' had indulged in very much the same sort of thoughts when they were in lower positions—either in Murphy Radio or outside. If that is the case, I assessed that the people concerned would be very quick to appreciate the misgivings of those below them, and would probably realise, too, that they might be in the best position to provide a remedy.

I must confess my own inability to suggest just what steps the 'office-bosses and charge-hands' should take, but I still feel that it may lie most within their power—by something they can say or do—to allay the fears that apparently exist. For my own part, I know that if I (just to take one example) were a charge-hand in the factory I should feel very seriously disturbed if I thought that even the vaguest gossip on these lines was floating about unchecked.

FRANK MURPHY."

Clearly, a number of the employees were reassured that no reprisals would be taken against them, for a variety of letters followed from individuals in both works and staff departments. Many of them pointed out the need for some mechanism for ventilating problems and bright ideas—some being in favour of a Staff Association, others of a kind of Works Council, and yet others for a traditional Trade Union. A number of people argued for shorter working hours (for their own department, alleged to be doing very tiring mental or physical work). When Frank Murphy was asked for his opinion, he pointed out that (1) workers doing different types of work had different hours; (2) if individual reaction to strain arising from type of work were to determine the hours of work, Murphy Radio would have to agree to 900 different hours of duty; (3) present hours of work were largely a matter of tradition and convenience; and finally (4) as far as he was concerned, a more fundamental question arose, viz.—was a human being a machine? Until someone answered that, he couldn't be interested in th subject of working hours.

Having given a fair amount of space to the bubble of independent thought now rising from the factory and laboratory, the editor endeavoured to reflect with equal fidelity the interests of the dealers. Thus in the same issue of June 27th 1936, an article on "The Finish of Murphy Cabinets" not only explained why Murphy sets were to be viewed as high-quality furniture in their own right, but also described the polishing process in some considerable and fascinating detail, no doubt opening a number of dealers' eyes to the care and expertise devoted to the appearance of the cabinets.

The article continued with advice on how to care for the cabinet. If the customer wanted a higher gloss, he was recommended to use a hard wax polish known as "Cerrax". French polish over the cellulose finish was never to be used; cracking might occur, with the resultant need to strip and re-spray the cabinet completely.

A brief but informative run-down on the progress of sales disclosed some puzzling variations in 1936 from the usual yearly pattern, and while frankly admitting that sales in May seemed abnormally low, this fact was pointed out, so that dealers could be reassured that it was a general experience, and could not be caused by individual mismanagement or slackness.

Another practical two-page article followed: "The Problems of the Second-hand Set". This distinguished between re-sale prices (where Murphy Radio gave guidance, subject to ruling market conditions) and allowance or trade-in prices, which "depend very largely on the individual dealer's technique—or lack of it". The article showed how to arrive at one price from the other, allowing for handling re-conditioning and selling costs, to proved a gross margin of 27½% on the sale of both new and second-hand sets.

The need to "stop and think about it" was stressed in the next article, written by Bernard Avery, a Murphy dealer from northwest London. After paying tribute to previous contributors from several progressive firms, he wryly commented on the practice of the "non-thinking" dealer who, having made a sale, left anything further to the service engineer, "who, as like as not, is also the person in charge of accumulator charging, maintenance of the van, and is electrician-in-chief to the shop, with standing instructions to sweep through the premises and dust the sets every time he finds himself without a job". This unfortunate fellow, said Avery, was given the task of installing the set in the customer's home, and if the aerial was inadequate or nearby electrical equipment caused interference with the set's reproduction, he was not instructed by his boss to do more than shrug his shoulders, whereas the installation of a decent aerial and the fitting of suppressors could produce a satisfied customer, and further sales resulting from his recommendations.

The conscientious service engineer was recommended to read a "Wireless Servicing Manual", written by one of the contributors to the *Wireless World* Journal, and reviewed by Geoffrey Baker, a member of the Murphy Laboratory Staff.

The issue concluded with thirteen pages describing and illustrating Murphy Radio's Annual Sports Day, which had by this time become a town-wide "family" event, obviously enjoyed by all participants and spectators. The editorial staff must have contemplated this issue with some satisfaction—surely they had managed this time to include something for everyone?

The next issue (July 11th) predictably showed a healthy response to Frank Murphy's question, "Is man a machine?" One reply was from his eldest son, Kenneth, who had worked for several years as a representative of the company, and who modestly signed himself "D. Kenneth" to indicate that his views were his own.

In fact, such was the general keenness to embark on lengthy philosophical arguments that the editor had to announce:—

> "WE'RE GLAD that the staff are availing themselves of the privilege of Free Speech. It's what's been wanted for a long time. BUT . . . we should be very grateful if correspondents would keep their letters as short as possible. Unlimited space is not available . . . 'Few were his words, but wonderfully clear.'—HOMER."

The office staff who had requested shorter hours because of the intolerable strain of working a 46¼ hour, 5-day week received a broadside from a long-suffering dealer:—

> "I can't refrain any longer! The poor slave-driven folks in the Murphy 'brain-works' departments are hard done by. Having to work from &.30 a.m. to 6.10 p.m. is more than most tough guys could stand, I am sure, but I wish some of them could come and exchange places with me for a few weeks. We here (boss included) work from 9.00 to 7 p.m.

(and 8 p.m. Fridays and Saturdays), and *no* Saturday off. Incidentally, 7 p.m. is a purely fictitious 'shut shop'—it's usually 10 p.m. before the last demonstration is done, anyhow from August to December. And there's no 'miking' when out on a dem. One's brain has to be on the 'qui vive' all the time, or bang goes a sale if one says the wrong thing to Grannie in the corner.

Don't think I'm complaining—I'm not. I enjoy it. But I just want to point out to the Welwyn 'brainworkers' that there are other brainworkers in the Murphy organisation (in its wider sense) who don't have half such cushy hours as they do . . ."

Two other subjects were also strong issues with dealers; one was the fact that the current Murphy sets were marked only in wavelengths and not with the station names, and the other was the lack of an all-wave model in the Murphy range. In both these matters it was felt that Murphy Radio were losing out against their competitors, and (even worse) disappointing the buying public. The Chief Engineer had previously replied in a rather Olympian way that there were a lot of snags to putting station names on the tuning dial, and until Murphy Radio could provide this facility with precise accuracy, they would not be embarking upon it. For customers' convenience, a separate list of station names with relevant wavelengths was supplied. In the following year, however, Murphy Radio produced a genuinely remarkable tuning scale on which station names were listed alphabetically and the set "tuned itself"—see Chapter Six). At this time short-wave sets were finding some favour with the public, and a courteous inquiry of "Why not?" from the owner of a Murphy set produced a reply from Frank Murphy himself. He contended that the firm hadn't yet brought out such a model because it wasn't convinced that the rather dubious advantages outweighed the extra expense. Experience with existing all-wave sets (remember, this was in 1936) had shown that, so far, all they offered was a greater number of stations broadcasting indifferent material accompanied by a large amount of background noise. "I think our view is that if we ever incorporate the short-waveband, it will probably in the first instance be only in the most expensive instruments we make, on the theory that to that type of buyer the extra cost is immaterial."

A more formidable onslaught was received from Dealer Birtles, from Sheffield. In blunt Yorkshire fashion he pulled no punches, and announced:—

"When are we dealers going to get more out of this Murphy business? Only when Murphy Radio realise that all good things come through the dealers. That will be a 'Grand and Glorious Day'."

He appeared to have no use for free speech (equated by him with "grousing"); he thought Murphy Radio had made a big mistake by omitting station names; he thought the standard of cabinet design had gone down from "the old days"; he was not surprised at sales going down; and as for Sports Day—he couldn't care less!

It says much for the courage of the Editor that he not only headed this letter "DEALER HITS US FOR 6!" but devoted his entire editorial to it, regarding it as a serious indictment of Murphy Radio policy. He asked whether other dealers agreed with Mr. Birtles that the firm was not "responding correctly to its environment"—pointing out, however, that this environment consisted not merely of *some* dealers, but of all the dealers, and also the suppliers, staff and the public.

But it was the centre page article that made the dealers blink. It was headed "DISTRIBUTION—AN IMPORTANT ANNOUNCEMENT FROM THE MANAGING DIRECTOR." In this they read that Mr. E. W. Kent had vacated the office of Distribution Manager on July 3rd and that Frank Murphy was taking over as Distribution Manager himself and had already held a series of meetings with all Distribution staff e.g. Area Managers, Representatives and senior inside staff. This would be "re-stating the Company's policy in a clearer way to both staff and dealers".

The resultant extra work held up publication of Frank Murphy's booklet on "Retailing", which they had all been exhorted to read and comment on, and since that had promised to be tough going, it was a relief to dealers to know that the booklet was delayed.

However, a second look at the paragraph might not have been so re-assuring. What had happened to Ernie Kent, one of the oldest and most respected members of the management? He was known personally to many of the dealers, and this was disquieting news. The proposed re-statement of Company policy might bring up again the dreaded subject of "Murphy Branches" as opposed to Murphy Dealers, which could easily split the unity of the organisation. What other disconcerting subject was Frank Murphy going to tackle now?

Pages 12 and 13 brought the answer. They were headed "GOVERNMENT IN INDUSTRY". These personal musings by Frank Murphy were introduced by the editor with some misgivings (obviously he was under fire from many people for allowing the *News* to become "too theoretical"), but he clearly thought it was his editorial function to provide a platform for all views:—

"On the Problem of Human Relationships.

Editor's Note. This article evolved from some notes which Mr. Murphy prepared in order to facilitate discussion of the subject between himself and several other learned people. I felt very keen to publish it as a reflection of Mr. Murphy's interest in the subject, even though, from the point of view of the *Murphy News*, it reveals a characteristic flaw—namely, that it starts off on a healthily practical note and then develops into the consideration of a deep abstraction.

I say this is a 'flaw' from the *Murphy News* point of view, because we haven't all got Mr. Murphy's capacity for wrestling with abstractions—most of us like something practical to bite on. At the same time, one must admit that few of the desirable concrete things of life would have developed except on a background of fundamental theory; hence I decided to publish the article, in the belief that this note of explanation would encourage other people to read it and think about it . . ."

Willby was certainly right. The article *did* wander off into abstract concepts, but it had a very practical bearing on the selection of both charge-hands and Murphy dealers, in pointing out their need to excel not only in technical competence but also in "human relations" ability. This concept, not exactly new to us now, was startling at the time. Frank Murphy went on to speculate how one could measure human relations ability, suggesting that one man might have high ability in the appreciation of beauty, another of humour, concluding with the characteristic sentence:—

"I may be on the wrong track in these thoughts, but it

seemed to me that in the ultimate they might have a very important bearing on such questions as Staff Associations, hours of work, 'Is man a machine?' and so on. It therefore seemed worth-while to offer them to the *Murphy News* so that other people could learn what was in my mind, and, if they felt like it, do their own spot of thinking on the subject".

He had not, however, concluded. A tail-piece, entitled "Another Big Question" raised another hare, which he was to chase for the rest of his life:—

"There is another subject that is biting me. I find that I want to state, almost as though it were an axiom, that every society, whatever its nature, must have a leader.

If that is true, what is the definition of a leader, and how does one know that any particular person is a leader?

And, of course, there is the other aspect of the subject; if it is true that every society needs a leader, how does it get its leader? Does it get it via the route of democracy (i.e. selection of superiors by inferiors) or in accordance with Bernard Shaw's statement that leaders are self-elected?"

Not exactly a typical article to find in a firm's house magazine!

Three other items in this same issue are worth noting. One, under the "Free Speech" label, is a discussion about the aims and possible adverse effects of introducing time-study engineers into a factory, between two well-known members of No. 1 Shop, Messrs. J. G. Naz and R. Nunns; the second is a brief factual report on the spontaneous growths of dealers' associations in Newcastle, West Yorkshire, and Scotland, from which it was obvious that dealers were reaping the advantages of thinking in depth about their work; and the third, most interesting from a sociological point of view, was a fierce discussion on the value and practicability of pacifism—one must remember that this was some two years before Munich.

A fortnight later, the next *Murphy News* showed signs of bursting apart, with the attempt to accommodate items as diverse as "Some Notes on the Operation of an Electrical Gramophone", "Service Glimpse No. 7", "The Need for Station-Names" (from a dealer), and "Man Needs an Ideal of Life" by Sir Richard Livingstone, "The Cause of Industrial Strife" (extract from R. H. Tawney's book, "The Acquisitive Society"), and "Analysing Human Relationships" by J. G. Naz, not to mention a report of the meetings of "The Ventilating Society", a new open forum established for all Murphy Radio employees without distinction of rank.

The editor manfully defended himself and his staff from dealers' criticisms that in recent months the *News* had become more and more of a staff magazine and less of a dealer's journal. He pointed out that "whereas at the beginning of 1936 the *News* contained items of roughly 90% 'dealer interest' and 10% 'staff interest' the balance was now 40% 'general', 35% 'dealer' and 25% 'staff' interest. Items like the staff correspondence on the sick pay question and the proposal to form a Staff Association were included in the 'general' section, because it was felt that everyone connected with the firm would be interested not only in the details of such subjects, but in *the distinctly unconventional experiment of giving people freedom to say what they thought*" (my italics).... "The *Murphy News* was never intended to be a solely-dealer paper. Its primary function was not to talk about the virtues of Murphy Radio and its products, but to discuss the sort of problems in which we are all fundamentally interested... ." Moreover, the works staff were not all seething with discontent, as some dealers had wrongly assumed... "Obviously, there are details which they would like to see improved, and given the privilege of speaking their minds, they would be less than human if they did not take the opportunity of raising such points. And after all, dealers themselves have set an excellent example in the use of the 'free speech' mechanism!"

In his editorial, Stan Willby remarked; "Six years cover roughly half the lifetime of the radio industry, and the entire manufacturing life of Murphy Radio Ltd.... The significance of all this lies in the fact that Murphy Radio Ltd., commenced manufacturing radio sets in July, 1930. The sale (note the singular!) for that month was ... one *B4*! July this year (1936) unless something miraculous happens next week, will have seen the despatch of something like 6,500 sets all told".

To celebrate this progress the front and back covers of the *Murphy News* showed a diagrammatic coloured sketch of "that part of Welwyn Garden City now embellished with buildings devoted to the designing, making and selling of Murphy sets, with a border of the various years, each with its appropriate models. The editor pointed out that "the first real factory which the firm had in 1930 is the building now marked 'Machine Shop'—that is, the right-hand section of the pair at the top end of Broadwater Road". (See illustration, fig. 22).

"On a more serious note, it can be said that Murphy Radio are very proud of the progress which the dealers and the public have enabled them to make in such a relatively short time. Although still be no means 'big', as manufacturing businesses go in these days, the Company enjoys a yearly turnover well in excess of a million pounds; it can claim, in all modesty, a very useful measure of public recognition and goodwill; and last—but definitely not least—it can claim the interest, support, and even friendship, of as good a body of dealers as any manufacturer ever had.

"To everyone connected with that achievement may we say 'Thank you'—and we hope to continue deserving it?"

The glowing moment of self-congratulation did not last long. The message on the cover of the next issue (August 8th, 1936) And now . . . the *next* six years . . ." indicated the typical Murphy "coats-off" attitude to the job. But a very serious editorial announced to its readers that the sixth anniversary marked the end of one phase in the Company's history in a very special way—because "*Mr. Frank Murphy has decided to free himself from the executive work of Murphy Radio in order to devote more time to other important activities which he has in hand. He has therefore relinquished the position of Managing Director, while retaining the Chairmanship of the Board. Mr. C. R. Casson and Mr. F. J. Osborne (later Sir Frederick) have been added to the Board of Directors and Mr. E. J. Power has become Managing Director*".

There is little doubt that this announcement was as big a bombshell to the editor as to many others in Murphy Radio. Probably only Frank Murphy's wife and the Board of Directors knew that although Frank Murphy valued the enthusiasm and spirit of comradeship which pervaded so many people now connected with the Company, his mind was now occupied with other considerations. The questions of the last few months:—"Is man a machine?" "What makes a man a leader?" "How do we select him?" "How does man fulfil his individuality!" were now of far greater importance to him than the efficient making and distribution of radio sets, and certainly far more important than making money. So he added a postscript to the formal Board announcement:—

"I feel that I would like to add something to this very bald statement, and try to indicate what the 'other important duties' are.

"It must have been evident to everyone that for some time past I have felt it important that we should have 'ends' to life, and to try and state what appear to me as worth-while ends. To me the subject continually increases in importance —as I believe that if I could only get the time to study it, and to talk to people about it, I might be able to produce reults that will help others quite a lot.

"I feel fortunate, therefore, that a practical re-arrangement of the work concerned in actually carrying on the Company's activities is now possible and that as a result I can, with a clear conscience, devote myself more fully to the job of discovering worth-while ends for us all to pursue.

"Needless to say, I feel a deep sense of responsibility to my colleagues, on account of the fact that they have agreed to free me in this way for a job which, after all, when judged by traditional standards, must appear very nebulous in character.

"Whether or not I shall succeed in digging out better 'ends' for us to pursue, time alone can prove. All I can say now is that I shall do my best.—FRANK MURPHY"

Four more pages followed this extraordinary declaration, all devoted to the problem of human relations, with contributions by R. H. Owen (No. 1 Shop), A. M. Kille (Chatham radio dealer), R. W. Duck (Cardiff radio dealer), P. M. MacNamara (Despatch) and J. G. Naz (No. 1 Shop); each thoughtful and provocative, and by no means suitable for casual reading. The final letter by H. R. Harris of London N.12. provides an enigmatic summing up of both the ccorrespondence and the editor's dilemma:—

"Frank Murphy asks, regarding the machine and men, 'For what use were they designed?'

Who can tell you for what purpose the machine was designed better than the designer himself? Who can tell you better for what purpose man was designed than his creator or designer?

But are the columns of the *Murphy News* for such questions

and answers?"

Editor's Comment: "The best debaters frequently answer one question by asking another. Is this question the final answer to Mr. Murphy's original problem?—S.G.W."

A sucker for punishment, Willby had also printed the long-promised article on Pacifism, entitled in bold black letters, "PEACE OR WAR?" by "The Newsman" (a pseudonym which disguised the identity of assistant editor Jack Hum). A forthright statement of the aims and implications of pacifism concludes:—

"It is a pessimistic outlook. It will remain so while this country prefers to out-arm the others rather than seek to remove the causes of war. I cannot foresee the disappearance of those causes until the nations realise that life exists to build an ultimately perfect civilisation, and not to be extinguished before it has had the time to realise its opportunities. Not until the world possesses men of sufficiently broad outlook to appreciate this truth and interpret it in no spirit of petty nationalism—not until then, will the world attain perfect peace."

Of course this article could not go unchallenged. The writer was labelled a "dangerous crank" by A. R. Turner of the General Office (himself a most gentle and lovable person). Dealer J. Mack of Llandudno and J. Lawson Trapp of London, N.8. upheld "The Newsman" from the standpoint of Christian principles, while Dealer Adcock of Ipswich and J. Greenberg, though agreeing with him, took a rather cynical view of the outcome for a pacifist Britain if it fell under the heel of an invader.

Understandably, a great many of the *Murphy News* readers thought that all this space devoted to issues like war and peace was misguided and quite irrelevant to their own lives, but the more thoughtful among them must have received quite a jolt from a long account by Mr. R. J. Butterworth, one of the partners in Murphy Radio's accountants, of the frightening adventures he and his wife had on a holiday in Spain. Innocently motoring through from France, they suddenly found themselves in the middle of the Spanish Civil War, and were very nearly killed out of hand by the Republicans (called Communists by Mr. Butterworth). Yet even this terrifying situation did not cause him to lose coolness or initiative, and ultimately he and his wife managed to return to France still marvelling at the contrasts in the Spanish character:—

> "The demeanour of the working people is forbidding to a degree. They are unshaven and morose, secretive and suspicious, and for ever plotting. On the other hand both Communist and Fascist at heart are just like one's own people, but even more willing to help or to give any service they can. Even the Communists who seemed (and were) such desperadoes on Monday were extremely lovable when you got to know them later on . . ."

But, of course, such a state of affairs could never happen in Britain—wars were only for excitable Latins. (Or were they? Oswald Mosley's Blackshirts were proof that Fascism could become a menace even here).

However, for the average Murphy dealer, the autumn sales rush was just starting, and most of them had neither time nor energy left at the end of the day's (and evening's) work to consider subjects of deep philosophical importance. The editor of the *Murphy News* evidently conceded that perhaps the balance had been tipped a little too far in the direction of philosophical rather than practical items, so on September 19th 1936, he had a cover design (with two dartboards) labelled "Fifty-fifty!" Inside, though, he admitted that philosophy would be taking a minor place in this issue after all, because space was needed for important statements on Murphy policy.

The first of these concerned the trade discount. It was announced that as from the new year, when the new models were introduced, the current discount to Murphy dealers would be increased from 27½% to 30%—something which many dealers had long felt necessary. The Company pointed out that "Although the long and careful investigations by our accountants have failed to produce conclusive evidence that the present discount is inadequate, it is nevertheless felt that in many cases the 'shoe is inclined to pinch'."

Some people might have thought the change came significantly soon after Frank Murphy's relinquishment of the post of Managing Director, but he himself in earlier days had said 27½% was not sacrosanct.

The second important announcement offered a new opportunity to dealers—a hire-purchase plan to enable the public to buy Murphy sets on weekly terms.

A third article told dealers that under a new scheme with a firm specialising in house-to-house distribution, they could gain direct contact with 5,000 possible customers at a cost of £6 per month. The idea was to make contact with people who did not see the dealer's advertisements in his local newspaper.

The reason for this practical boost towards dealers' sales efficiency is clear when one reads a remarkable analysis by P. K. O'Brien, the company's economist, on "The Progress of Sales". Referring back to an article earlier that year, when dealers had been told frankly of a puzzling drop in sales in May (when, in theory, the new models should have boosted sales) O'Brien now concluded, after carefully taking into account a number of relevant factors, that June, July and August were also below estimate because (1) fewer new dealers had been recruited and less turnover resulted than was expected, (2) there had been some transfer of sales by larger dealers from Murphy to non-Murphy sets. (In fact, to Bush Radio—see Chapter Seven). Most firms, he admits, would be pretty satisfied with an increase of 30% in units (sets) sold and 15% in turnover in one year, but that was not good enough for Murphy Radio . . .

Quite apart from the intrinsic interest of the figures, and the fact that as early as 1936 the company was monitoring its own progress as scientifically as possible, there is surely some significance in its willingness to make them publicly known through the medium of the *Murphy News*. Indeed, editorial comment (October 3rd, 1936) points out that the *Company practice of making known turnover and sales figures, profit policy, factory output, employment statistics and similar subjects was very much contrary to normal industrial procedure.* There were obvious risks in such openness, but generally speaking the firm felt its policy was wise. The words of the editor echo Frank Murphy's thinking:—"In practice, it seems inevitable that, if individuals are to be given the opportunity of full development, they must always be offered responsibility a little in advance of their experience. Otherwise, how is the experience to be gained? Obviously, such a course involves risks, but nothing that is worth having is achieved without them".

Meanwhile, what was Frank Murphy himself doing? Though no longer playing an active part in the day to day running of

JANUARY 11, 1936

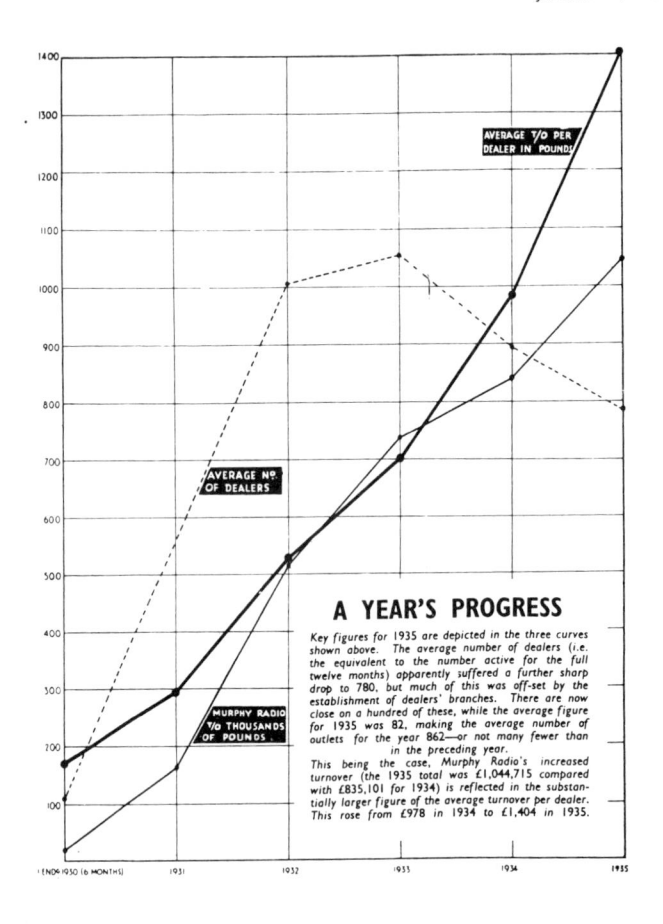

A YEAR'S PROGRESS

Key figures for 1935 are depicted in the three curves shown above. The average number of dealers (i.e. the equivalent to the number active for the full twelve months) apparently suffered a further sharp drop to 780, but much of this was off-set by the establishment of dealers' branches. There are now close on a hundred of these, while the average figure for 1935 was 82, making the average number of outlets for the year 862—or not many fewer than in the preceding year.

This being the case, Murphy Radio's increased turnover (the 1935 total was £1,044,715 compared with £835,101 for 1934) is reflected in the substantially larger figure of the average turnover per dealer. This rose from £978 in 1934 to £1,404 in 1935.

Murphy Radio, he was certainly not idle. He was keeping up a steady stream of correspondence, interspersed with personal interviews, with various eminent University dons, among them Sir Richard Livingstone, author of "Greek Ideals in Modern Life", whose enthusiasm for the Danish Folk High Schools had been quoted in an earlier issue of the *Murphy News*. He was also doing a prodigious amount of reading, particularly of Maine's "Ancient Law" and other weighty tomes, in an attempt to define man's "rights" and "responsibilities". Some of his Cambridge friends had colleagues in the United States, who, they felt, would be interested to exchange views with him. Accordingly on September 19th 1936, he sailed on the 'Empress of Britain' for America, taking with him his driver, Alfred Hood, and his well-known Rolls-Bentley motor car. There were pictures of "The Guv'nor" and E. J. Power just before the ship sailed, in cheery mood, despite the characteristically heavy schedule which had been planned.

The trip was intended to last about twelve weeks, but in fact Frank Murphy was home again by early November, as was announced in the *News*. "He arrived in the Garden City on Tuesday afternoon, having covered 6,000 miles of American roads in the Bentley, met and talked (you bet!) to close on seventy leading Americans in business and the universities, and experienced the worst Atlantic crossing the Queen Mary had yet had to face—and all in a matter of seven weeks or so!

"At the moment of going to press, we've only seen him for a matter of minutes, but we can reassure those interested that
   (a) he still looks the same
   (b) he still smokes a pipe and doesn't chew gum
   (c) he still sports a cloth cap and a walking stick
   (d) he still knows what to do with a cupper tea."

A full account of his tour was promised in a later issue, and in fact was printed in two halves (on November 28th and December 12th 1936), while an additional article appeared on October 31st written by Frank's sister, Mrs. Ethel Karsay, who came back to England with him, and in the last issue of the year, Alfred Hood, the driver, contributed some impressions of his own about the trip.

With the first issue of 1937, the editor looked back at the past year with few regrets. Undoubtedly, 1936 had been a tough year for many of the dealers, who had not hesitated to pass on their customers' complaints to the *News*; no station-names on sets, minority-taste cabinet design, no all-wave set in the range, none of the accustomed Murphy "firsts" in radio design which had sustained them as leaders of the industry for the past three years, unsettling discussions over distribution and fears of being taken over as Murphy branches"; the resignation of Frank Murphy as Managing Director—all these had combined to depress their spirits.

Still, there was quite a bit on the credit side, when the year was examined in detail. First, it was not true to say there was no break-through in design. The year 1936 saw the very successful introduction of the *B23* model, a battery set selling at the incredibly low price of £6 7s. 6d., and demonstrating Murphy Radio's policy of steadily widening their range to bring good radio within the reach of lower-paid consumers.

Secondly, a real effort had been made to try to even out the peaks and troughs of production and employment at the works, by introducing the *"Stocking Plan"* which asked dealers to order sets regularly throughout the year, even if they could not immediately sell them, and the *"Even Load Plan"* of regular production of sets in the factory evenly throughout the year to reduce the necessity to lay off workers in slack selling periods (although a hundred or so were regretfully discharged in the autumn). Still, regular hours and regular pay had been enjoyed on a scale not hitherto experienced in Murphy Radio, and most certainly nowhere else in the industry. (See Appendix 2).

The staff had also benefited by the Company's initiation of payment for holidays and sick pay, and they had not only expressed their gratitude for this, but shown that they wanted to convey their ideas on management and working conditions through a Staff Association, or some similar type of organisation. They also enjoyed, by now, the use of a new brick-built canteen and clubroom, with scope for many activities outside work.

Several events had occurred in 1936 which were portents for the future. The dealers' trade discount, so long a bone of contention, was to be increased to 30% as from the new year; the number of dealers was still to be based on the "one per shopping centre" theory, but this was to be interpreted more flexibly, especially in the bigger towns and cities; three new *Murphy Dealers' Associations* had been formed; and road transport of sets had become more and more common. Managing Director E. J. Power had addressed his first series of dealers' meetings in October, and his factual approach to the Company's programme for 1937 no doubt reassured those dealers bewildered by Frank Murphy's increasing pre-occupation with the theory of retailing and, indeed, with the purpose of human existence.

Best of all, from the dealers' point of view, 1937 opened with a tremendous fillip: the introduction by Murphy Radio of station names on the tuning scale—and not just a jumble of names, but names arranged alphabetically, which was a technical advance of a very high order. Moreover, this alphabetical scale was fitted not just to the most expensive set in the 1937 range, but to all of them, except the very cheapest battery set. The illustrations of the new models showed that the cabinet designers also had taken to heart the previous year's criticisms, and the new *B31* £6 10s. 0d. battery model delighted dealers' hearts with its simple highly-polished walnut frame, black sides and tuning knobs, and a perfectly circular opening (not a "Chinese circle" for the loudspeaker, which had been unpopular in the 1936 range of models). Similarly, the *A34* table model, retailing at £11 10s. 0d., had a very simple design in mottled mahogany, again with a highly polished finish.

Such was the exhilaration of dealers, particularly over the *B31*, that within three weeks the Company had received orders for 6,000 sets for immediate delivery. One letter is typical:—

"This afternoon, with great hope in my heart, I accepted delivery of the first of the Murphy range for the present year—namely, the *B31*. Let me compliment you on turning out a really wonderful receiver. I have given it only a rough test so far, but I am delighted with both the results and the general appearance and finish of the set.

This is the cheapest and, supposedly, the 'poorest' of your range this season. If so, then I hope the brandy is near when the others arrive!"   F. A. SWAIN (A. P. Merriott, Bristol)

All the issues of the early months of 1937 reflected this new-kindled enthusiasm, as dealer after dealer wrote in to express satisfaction with the new models, with only minimal criticism on minor points—very different from 1936. At the same time, the dealers were paid the implied compliment by the editor of being capable of sustaining a theoretical discussion, because in each issue he printed detailed reports of current BBC discussions on the subject of distribution.

A series of articles on Consumer Research by the Murphy economist P. K. O'Brien broke new ground. O'Brien also contributed four articles on "Territorial Research" which covered, first, the definition of a shopping centre; next, how the dealer could discover the strongest and weakest parts of his territory; and finally how the dealer could expand his business, given this knowledge.

It is remarkable to reflect just how far the services offered by Murphy Radio to its dealers had extended since the early days of simple publicity aids. The popular series of "Candid Comments" on dealers' window displays was resumed, in response to many requests, and there was a steady increase in the number of technical articles on television, then in its infancy at Alexandra Palace.

But there was one major item of news which escaped publication in the *Murphy News*. Stan Willby, the Editor, was not present at the fateful Board Meeting in January 1937 when Frank Murphy, realising that the other Directors felt that the Company would now be better off without him, voluntarily resigned as Chairman, and announced that he was severing his connections with Murphy Radio.

The news spread like wildfire round the works and caused enormous speculation. None of the Directors made any official statement at the time, but recently Rupert Casson explained that for several months Frank Murphy had become increasingly sceptical about the possibility of serving the public within the framework of a limited liability company, whose terms of reference were primarily the making of maximum profits. He bewildered the other Directors by playing devil's advocate, and suggesting all kinds of outrageous policies totally at variance with his own principles of integrity and service. Perhaps he thought that someone would call his bluff and challenge his deliberately specious arguments. But no one did, and the meetings ended in exasperated stalemate, until finally he said, "I suppose you think the Company would be better off without me!" There was a silence, and then Rupert Casson said slowly, "Yes, Frank, I think it would". Whereupon Frank immediately announced his resignation, and gathered up his papers.

Three weeks later, Stan Willby wrote in the *News* that he had been informed by Frank Murphy that he was proposing to tackle the retail distribution of furniture as his next job. The editorial expresses both admiration and apprehension:

"It will come as no surprise to those who know Frank Murphy that his future activities are likely to be in the direction which has absorbed so much of his energy and interest over the past year or two—namely, retail distribution. Most people know his strong conviction that normal distribution methods tend to increase prices and consequently keep down the standard of living. With that viewpoint, it is natural that Frank Murphy should select a major 'necessity' trade as the scene of his new venture, since in that direction lies the greatest possibility of achieving his underlying aim."

"He has certainly chosen a stiff furrow to plough, but we know that everyone in and connected with Murphy Radio will wish him the best of luck."

"Even if we don't see him as often in the future, we shall be tremendously interested in watching his experiment. One thing we are certain of is that, although he is bound to strike an unusual note in his new sphere, and other people may find it a bit uncomfortable at first, in the end many are likely to benefit by his efforts."

After that, there was no other reference in the *Murphy News* to the activities of the founder of the company, though the *Murphy News* itself continued to be issued up to 1962, the year when the Company was taken over by the Rank Organisation. E. J. Power's face had replaced Frank Murphy's in the national advertisements by September 1936, and Dr. R. C. G. Williams had been made Chief Engineer of the company. Gradually the famous "Man with the Pipe" logo disappeared from dealers' windows and cartons, being replaced by the phrase "Your Murphy Dealer". No credit was given to Frank Murphy as the originator of the Rent Theory, when this occasionally appeared in O'Brien's articles on Territorial Research. He had become not so much invisible, as unmentionable, though many of his old dealer colleagues discussed his post-radio activities, and some actively supported him.

I think it is fair to say that the *Murphy News* never again recaptured the innocent exhilaration of those heady early days; it became a good, readable, but predictable publication, and its interest for the non-technical reader sharply diminished.

When the war came in 1939, Murphy Radio had to switch its output to contracts for radio and radar equipment of the armed forces. Like other companies, they were restricted by the Government to producing a single model of uniform design for civilian use.

By this time the chief executive—in fact, if not in name—was Dr. Williams, now Professor R. C. G. Williams. He recalls an amusing incident when a certain George Brown helped him to solve a rather tricky situation. There ws a strike at Murphy's—a most unusual event—over an extra ½d. per hour for the workers. The government contract gave Dr. Williams no room for manoeuvre, and the Murphy workers were hampered by not being represented by a trade union. The local Branch Organiser for the Transport and General Workers' Union at that time was George Brown (later Lord George Brown). Rapidly sizing up the situation, he got permission from Dr. Williams to interview the workers, signed most of them up to join his union, and thus provided a suitable negotiating body with whom the management could do a deal!

Today, after a distinguished academic and industrial career Professor Williams is a consultant of international standing in the electronics field. It is a long way from those pioneer days at Murphy Radio, yet he still recalls Frank Murphy with gratitude and admiration as "an idealist with a flair for human relations".

Meanwhile, after the war ended, the Company continued to produce fine radio sets and its profits continued to rise until the difficult years of the late fifties and early sixties, when the whole industry was staggering under the blows of Japanese competition.

In 1962 Murphy Radio was taken over by Rank Radio International, who had already acquired Bush Radio. The Managing Director, Dudley Saward, announced:

> "When a new range is introduced, the identity of Murphy is its presentation, which will be jealously guarded. It is our plan to preserve the Murphy identity."

This they certainly did, right up to the sad day in 1981 when the Rank Organisation decided it could no longer continue with its loss-making radio and electronics side. Yet in August 1978 the

Government had sponsored a £10 million deal between Rank and the Japanese company Toshiba, to set up a new joint venture in Plymouth, using Japanese technology. Rank Toshiba employed 3,000 people, making Bush, Murphy and Arena television sets. In 1979 *IDP News*, the successor to the *Murphy News*, produced a front-page article headed "50 YEARS OF MURPHY", with a nostalgic look back to the early days, with photographs and reminiscences provided by the old dealers and their sons.

Two years later, both Rank Toshiba and Rank Radio International were closed down, and the famous Murphy trade name was sold to *J. J. Silber*, a subsidiary of Great Universal Stores. Meanwhile, Toshiba re-opened part of the Plymouth factory, employing 300 people to manufacture a range of Japanese television sets for the British market.

Just one announcement about the new Toshiba venture would have pleased Frank Murphy. It was to have an Advisory Board elected by *all the work-force* (not just Union members), which would have access to *all Company information*, with the aim of ensuring genuine consultation with the shop-floor.

"Very nice!" no doubt he would have said. "But can the workers disagree with the management without fear of the sack?"

Chapter Nine

A NEW CONCEPTION OF BUSINESS

"I was thinking—which is the best way out of this wood!"
Alice Through the Looking Glass

Delving into the problems of efficient distribution had shown Frank Murphy that the intelligent use of mathematics could provide part of the answer; but even more important was a study of the motivation and natural instincts of the buyer and the seller. Consequently, while retaining a healthy respect for the economic viability of a business, he was inevitably turning more and more to consideration of human relations in industry, and how a more satisfying and secure life might be achieved for everyone, whether worker, director or shareholder.

1. Integrity—what is it? How important is it to us?
2. Equality—Can all men be equal in any respect? If not, is this an obstacle to a free society?
3. Do we need leaders in industry, or in society as a whole?
4. If we do, how should we obtain them? By election? or by self-selection?
5. What qualities should we take into account when appointing someone to fill a vacancy?
6. On what grounds should a worker be dismissed?
7. Can someone of integrity work for a company whose primary aim is to make maximum profits, i.e. by exploiting the consumer?
8. If we accept the principle in civil life that all must obey the Rule of Law, how can we apply it to industry?

In Murphy Radio, he had created an organisation with some two thousand individuals united by a single aim—the highest possible service to the consumer. Yet it was evident that many of these individuals were under severe strain from attempting to reconcile conflicting loyalties—for instance, dealers striving to make an honest livelihood within the confines of a strictly limited discount; workers keen to give their best, yet ever fearful of the annual possibility of redundancy; management anxious to allow workers their say, but not at the expense of meeting production targets; a board of directors understandably nervous of radical change which might set back profits.

Frank Murphy found himself asking whether the present capitalist system and its instrument the liability company could give to every worker the same right to justice as he was entitled to as a citizen of Great Britain. The answer did not appear to be promising. What seemed to be the case was that a fair-minded employer could run his company like a benevolent father-figure in prosperous times, but when there was a slump, or when greed for ever greater profits overcame his better judgment, he would ignore his workers' interests and put his own first. Naturally, the workers joined together to try to protect their interests, and struggles between unions and management frequently resulted in lengthy strikes or lock-outs which damaged everyone, including the consumer.

In civil life, after centuries of effort, the British had achieved a democratic constitution in which all citizens had equal rights under the law. By studying the history of how those rights had been achieved, it might be possible to formulate similar principles in industrial life, and so provide rights for all in the working environment.

It was these considerations which led Frank Murphy to plunge into a study of such classics as Maine's "Ancient Law", Bagehot's "English Constitution", Trevelyan's "History of England", Dicey's "Law of the Constitution" and Lindsay's "Essentials of Democracy". Going further afield, he began reading Barfield's "Law, Association and Trade Unions", Walter Lippman's "Good Society", Livingstone's "Greek Ideals and Modern Thought", Spender's "Government of Mankind", and Lin Yutang's "Importance of Living". He delved into semantics with Korzybski's "Science and Sanity", and compared Hitler's "Mein Kampf" with the Analects of Confucius, as guides to modern society. He had already been sufficiently impressed by a book called "Successful Living in this Machine Age" by an American businessman, Edward A. Filene, to send a free copy to every one of his dealers. He also suggested they had another look at Lewis Carroll's "Alice Through the Looking Glass".

His reading reminded Frank Murphy of the basic achievements of British democracy. These he summarised as follows:

Under the feudal system the serfs had the right to protection by their overlords, in return for unquestioning obedience and service, while the King in turn imposed his central authority over the barons and lords.

When Magna Carta was signed, the barons imposed the Rule of Law over the King, and judges were made independent of the monarchy. The jury system (the right to a free trial by one's equals) was established.

Under the Tudor kings, central authority returned, but the seventeenth century saw the emergence of Parliament as the dominating power, after the fierce struggles of the Civil War. Freedom of worship and freedom of speech for all citizens were eventually won during the next two centuries. With the rise of the trade union movement following the Industrial Revolution, freedom of association was conceded.

In th present century the campaign for civil rights continued with the suffragette movement, and the fight for married women to own property was another step forward. The right to free education for all children was granted, and gradually extended through secondary education to the universities. Basic assistance from the state for workers who were sick or unemployed was guaranteed by Act of Parliament. As many people from the New Commonwealth came to Britain to settle here and seek work, they and their children continued to struggle for equality of opportunity, and to be recognised as having civil rights equal to those of other British citizens.

The fight for individual liberty has been enshrined in many phrases, the most memorable being "No rights without obligations"—"No man to be judge in his own cause" and "All men to be equal before the Law". It seems that the greatest degree of freedom at all levels in society has been achieved when men have agreed to apply and re-interpret these three basic principles of the Rule of Law. On the other hand, dictators like Hitler, with their watchword, "Might is Right", were always encroaching on the freedom of others to maintain their position of power.

Now at last Frank Murphy felt he was getting near to the heart of the matter. *What industry needed, he thought, was a framework for the Rule of Law, with the worker's basic rights and obligations embodied in a Constitution,* so that freedom and security could be available to all. He envisaged an industrial organisation which would operate on a commercial basis but which, unlike the conventional limited liability company, would have as its agreed aim the intention *"to apply knowledge with integrity and to express individuality in the service of society".*

This simple-sounding phrase was not arrived at without many hours of thought, and discussion with valued friends, such as Arthur Turner, Frank's brother-in-law, whose knowledge of British history ws equalled by his love of his country and pride in her achievements, and who also had many years of experience of the commercial world, in banking and shipping. "Dear Old Art", as he was affectionately known to the family, could be relied on to sift the wheat from the chaff in Frank's ideas, and to point out gently where enthusiasm might have led beyond the bounds of sober reason. Although Frank delighted to tease Arthur Turner, he had an immense respect for him, and very much valued his opinions. So together they studied the structure of industry, and particularly that of the limited liability company. It is interesting and sadly ironic that what they had to say about the latter is still true today, fifty years later, and Britain's industrial troubles far from diminishing, are even more severe and intractable.

This is what they had to say:

> "The normal Limited Liability Company is controlled by a Board of Directors, with executive power in the hands of a Managing Director, and the shareholders who provide the financial resources have no say in the operating policy. Even at an Annual General Meeting, it is rare indeed for shareholders who disagree with company policy to vote for the removal of the Chairman and Board of Directors, and rarer still to achieve it. It is natural that since the majority of small shareholders are interested only in the company's ability to make profits and pay them dividends, they should be content to have no knowledge of the company's aims and methods, unless there have been disastrous losses, by which time it is generally too late for a new policy to be introduced. Similarly, in all companies except those run as workers' cooperatives, the workers have no direct say in company

policy, no knowledge of its financial position, and no security of employment if they are rash enough to criticise the management. Consequently their energies tend to be directed towards the one matter they can influence—namely, their wage-levels, and conditions of work.

The Managing Director has more security than his employees, but even he can be sacked by his directors for reasons quite other than his technical competence, and the Chairman and Directors can openly or secretly indulge in mutual in-fighting, which can destroy the company's viability."

What Frank Murphy argued was that if the primary aim of all these individuals was not to secure the greatest possible power or profit for themselves, but to apply their knowledge in the service of society (e.g. the consumer), they would achieve mutual co-operation, individual job-satisfaction, and in all probability a greater financial return than under the old system.

With this in mind, Frank Murphy proposed a re-shaping of the Limited Liability Company, which would give management, workers and shareholders each a meaningful role, yet allow the company to operate without infringing the regulations of Company Law.

Thus, instead of Directors, he proposed to have a *Board of Trustees* who would have no monetary interest in the Company, but would have the power to appoint and to dismiss the Managing Director, to authorise his salary, and to deal with any complaints brought against him. They would be in many ways similar to judges in the law courts, since their main function would be to uphold and interpret the constitution, or principles on which the Company operated.

The Managing Director (who was to be re-titled Governing Director), while having full executive powers, *would not be able to control the Company for his own financial gain.* He would be allotted sufficient vote-carrying but non-dividend-paying shares to comply with his statutory function at an Annual General Meeting, but he could not out-vote the combined votes of the Trustees and therefore could not alter the Articles of Association (or constitution) of the Company. All the rest of the shares would be ordinary dividend-carrying shares, without voting powers, available in units of £1 each, available to all except the Trustees and the Governing Director. These proposals would be acceptable in terms of Company Law.

The shareholders would be actively encouraged to take an interest in the company's progress, not merely by receiving an Annual Report by the Chairman, but through a monthly newsletter or magazine, in which they *could make their views known and join in discussions on policy.*

Probably the most fundamental change, however, proposed by Frank Murphy related to *the appointment and dismissal of employees.* One of the basic rights of every British citizen is his right to appeal in a court of law for his case to be held publicly before an impartial judge and a jury composed of his fellow-citizens. In the nineteen-thirties a worker could be deprived of his job, and therefore of his livelihood, by his immediate superior in the management line, with no right of appeal. Frank Murphy believed this to be unjust, and at the root of many bitter confrontations between management and labour. He therefore suggested transferring to the industrial situation some of the rights which people enjoy as citizens. For instance, where a vacancy occurred, the person in charge of that department should select someone whose character and ability seemed to him suitable and recommend him or her for appointment. This would be for a probationary period—say, six months—at the end of which the new employee would come before a panel of assessors, two of equal rank, two of rank above and two of rank below. If the panel were unanimous in accepting the employee's character as excellent, then he or she could not be dismissed thereafter except on a charge of *"lack of integrity"*, and then only through the unanimous decision of a similar panel of assessors. The accused person would have a right of appeal to the Governing Director, whose decision would be final.

But what exactly is "lack of integrity"? Frank Murphy and his friends found that, like the elephant, it was difficult to define, but easy to recognise. Eventually he included as No. 29 in the Articles of Association the following paragraph:—

> "Lack of integrity shall be considered to have been exhibited by any person who commits a deliberate act which is, or is intended to be, detrimental to the religious, political, technical or economic interests of the group. In the event of the group interests conflicting, the order given shall also

be their order of importance."

In a footnote, the reader was reminded that a similar practice already obtained in banks, insurance companies and the civil service, where dismissal was rare except for "grave irregularity"—or, in other words, "lack of integrity".

In civil life, our fundamental right to freedom of speech is acknowledged even by those in positions of power who find it inconvenient, embarrassing, or openly obstructive. In recent years some of the fiercest public debates have arisen over questions such as the right to organise protest marches, demonstrations and sit-in strikes in support of minority views. The freedom of journalists to report and comment on current government policy; the right of church leaders to express their views on nuclear disarmament and mass unemployment; the BBC's insistence on "balance" in programmes devoted to sensitive issues; these all show that the right to freedom of speech can never be taken for granted.

Frank Murphy's experiment at Murphy Radio had shown him the remarkable effect of giving people the freedom to speak their minds through the medium of the *Murphy News*. His decision to make public in this way the Company's future plans, and to admit unhesitatingly that errors had been and still could be made, far from ruining the dealers' trust, hd strengthened it. So Article 19 of the Constitution of the *New Conception of Business* laid down that

"The organ of free speech shall be a monthly magazine, in which shall appear all material information as to the progress of the Company."

This information would be conveyed not only in articles and reports but by the publication each month of a balance sheet, showing the financial position for the previous month, and also the previous year. Contributions to the magazine were to be welcomed from all, regardless of rank or reputation, and in Frank Murphy's opinion, if people were genuinely interested, they would be only too keen to do so. One has only to recall the shrewd and telling points contributed by members of the public in present day programmes like "Any Questions?" and "Open Door", to realise how right he was.

As might be expected, the New Conception of Business laid down regulations on the welfare of employees which were remarkably generous for its time. Various clauses in the Constitution cover the payment of non-contributory pensions for all full-time employees of ten years' standing; not less than two weeks' annual holiday with pay for all; full pay less National Health insurance benefit to any employee absent through unavoidable sickness, subject to a doctor's certificate on request; and an ex-gratia payment on marriage to all women employees leaving for this reason. All employees, including the Governing Director, were to be required to undergo a medical examination before being permanently appointed. All wages were to be on a flat rate payable weekly (including those of the Governing Director), and there was to be no payment for overtime, any hours in excess of schedule being regarded as a voluntary act on the part of the employee. (A footnote to this last clause states that "habitual overtime is occasioned by inefficient management, and is consequently avoidable.") A further comment envisages a kind of 'flexi-time', i.e. "some system of variable hours, with longer hours in the busy season and shorter in the slack, without varying rates of wages".

Lest prospective employees should be under the illusion that working for a company operating the New Conception of Business would be like basking in perpetual sunshine, Article 35 remarked bluntly:

"Since 'lack of integrity' is the sole ground for dismissal of any employee of the Company, subject to the condition previously stated (compulsory retirement due to ill-health), reduction in wages occasioned by economic necessity must be borne by the entire staff including the Governing Director. Every individual must be prepared to accept a reduction which is proportionate to his or her current wage or salary."

Here he was no doubt remembering the early days of Murphy Radio, when, in order to enable the new company to survive the National Depression and the competition of stronger rival companies, he had asked everyone (including himself) to accept a ten per cent cut in wages until the sales of sets began to climb again.

Twelve years later, in 1950, Frank Murphy was to write a full account of his philosophy, which he called *The Root of the Matter*. In it are the same phrases which inspired him throughout his life:—

"*The object of a business, like that of every human organ-*

isation, should be to enable us to express our individuality in the service of others.

This can be achieved in business by changing its basic belief or terms of reference for shareholders, directors and staff alike to:—

NO RIGHTS WITHOUT OBLIGATIONS:
NO MAN TO BE JUDGE IN HIS OWN CAUSE:
ALL MEN TO BE EQUAL BEFORE THE LAW—that is, to the Rule of Law."

Chapter Ten

GOOD FURNITURE FOR ALL.

"Nothing astonishes men so much as common sense and plain dealing." *Emerson*

Working out the principles of the New Conception of Business kept Frank Murphy fully occupied from January 1937, when he resigned from Murphy Radio, to June 1938, when he was ready to form his new company—Frank Murphy Ltd.—in which those principles were to be put into practice.

The Prospectus of "Frank Murphy Limited", issued on 14th June 1938, offered initially "10,000 six per cent cumulative participating shares at one pound each", and by September some £7,000 had been subscribed, £1,600 of which was earmarked to pay for office equipment, cars, salaries, publicity and general expenses. The approximate weekly expenditure was calculated at £70. Only 25 shareholders invested over £100 each—the great majority of holdings were of under £20. Many of these were taken by Murphy Radio dealers, keen to support their former "Guv'nor" in his new venture, hoping it would herald the dawn of genuine industrial democracy.

September 1938 saw the first issue of the *Murphy Review*, a monthly magazine intended to give information on the Company's policy and current progress and space for the exchange of views of all interested in the New Conception of Business. The *Review* began with photographs and brief biographies of the five Trustees of Frank Murphy Limited. These were *John Alcock* of Stoke-on-Trent, Murphy dealer in the Potteries with seven branch shops, a Methodist Lay Preacher and a keen believer in practical Christianity; The *Rev. Reginald Ernest Fenn*, formerly headmaster of a school in China, and now Minister of the Free (Presbyterian) Church in Welwyn Garden City; *Arthur Robinson Turner*, of Holland-on-Sea, formerly a shipping agent in London, and a lifelong friend of Frank Murphy; *Laurence Beddome Turner*, of Cambridge, a University Lecturer in Engineering, who first knew Frank Murphy before World War I when both were in the Post Office Research Department; and *Herbert George Wood* of Birmingham, Lecturer in History and New Testament Studies, who had written many books on Christianity in the modern world. Each Trustee made a brief statement of his reason for agreeing to stand as a Trustee of the Company.

Then followed twelve pages of varied statements by individual shareholders on "Why I invested". Most declared that they were in sympathy with the idea of a square deal for shareholder, employee and customer alike, and quite a number mentioned their admiration for Frank Murphy himself as a motive. The desire to make big profits was noticeably absent—in fact, several correspondents were quite prepared to lose their money in a good cause. Perhaps a typical example was that from a shareholder in Caterham:—

"My reason for investing my small amount of ten pounds in your venture, was because I wanted to do my little bit in furthering a project which is attempting to put ideals into practice. I showed your explanatory booklets to a number of my friends in the electrical industry who without exception agreed that you were either a clever knave or an honest social reformer. For myself I had to confess that these alternative definitions were in the circumstances permissible, for even the most enlightened employer of today is inclined to look upon his good deeds to his employees and customers as merely means to an end—bigger dividends—and there seemed to be no real reason why the methods you adopt are not in the same category. However, your convincing booklets won me over to the conclusion that your main motive was to revolutionise the whole modern outlook on business and to constitute not merely a "money-grabbing" organisation, but rather a brotherhood of mankind, whose sole purpose was service to the community and to each other. So I decided to stake my faith, and if necessary lose my savings in the altruistic enterprise of Frank Murphy Limited.

One of my friends asked me the quite natural question "Are you out after your fortune or the good of mankind?" My answer was, "Both, if possible", and presumably that is an answer after your own heart."

Two practical, down-to-earth articles in the first issue of the *Murphy Review* dealt with:

(1) The Company's objects, policy, organisation, and use of consumer research to establish the highest possible value for money ratio (unsigned, but certainly by Frank Murphy himself);

(2) *Retailing in general, and the Company's decision to open its own shops* (or rather showrooms) rather than use existing furniture outlets.

This second article was contributed by Kenneth Murphy, who had left Murphy Radio to join his father in the new venture. From his photograph (fig. 32) his eyes gaze steadily at the reader. His expression is serious, yet there is a hint of humour in the full sensitive mouth. But Kenneth was no cardboard hero. He had already, as a popular and hard-working representative in tough London districts, become a respected figure among colleagues in Murphy Radio, and he was now to contribute months of valuable work in building the nucleus of a distribution organisation for Frank Murphy Ltd. His early death in 1942 was a personal tragedy, since he never reached his full potential. At this moment, however, he was full of life and thoroughly enjoying his job and his new responsibilities as Deputy Governing Director.

The subject of furniture was approached by Frank Murphy not from the aesthetic, but the engineering point of view. His first article stated:

"It is convenient to divide furniture into five groups, namely:—

1. *Sitting apparatus* (anything from a stool to an armchair—primary requirement—comfort).

2. *Sleeping apparatus* (a bed, obviously—primary requirement—rest).

3. *Tables or benches* (a dining table is a table or bench—primary requirement—a rigid smooth surface supported at the right height for comfort when sitting).

4. *Looking apparatus* (a mirror—primary requirement—to see yourself without distortion).

5. *Filing apparatus* (wardrobes, chests of drawers, sideboards, bookcases—primary requirement—that the articles are conveniently contained in them and easily

accesible when needed).

A writing bureau is a combination of a table and a filing apparatus.

A dressing table is a looking apparatus in combination with a filing apparatus and a bench . . ."

To design these items, he goes on to say, it would be necessary to know in detail how they would be used. For instance, in the case of the dining table, how many people would it normally seat, and how many on special occasions? Would the children do their homework on it, would mother rest her sewing machine on it, would she do her ironing there? and so on.

Another interesting fact revealed by Frank Murphy was that in 1938 oak as a raw material cost approximately 2d. per lb. and that most oak tables, except those custom-made, worked out at 10d. per lb. ex works, thereby bearing out the general rule of many industries, that the manufactured cost was approximately five times the cost of the raw material (assuming a large production run).

This explains the otherwise mysterious front cover to the first *Murphy Review* which showed Frank Murphy seen gazing meditatively at the public through a pair of scales, with a weight in one pan and a miniature table in the other, labelled "10d. per lb.".

The final pages of this first issue of the *Murphy Review* contain brief sections labelled "Statistics", "General Finance", "Details of Staff" and "Book Notices", all of which were to be regular and highly informative features in subsequent issues, faithfully carrying out the principle embodied in the *New Conception of Business* that all relevant information must be made freely available to shareholders, staff and consumers, but particularly to the former, so that they could keep a regular check on the aims and progress of the Company in which they had invested.

In the statistics section the reader learns the result of one of the preliminary questionnaires, in which consumers were invited to state their annual income, and the price they would consider reasonable to pay for individual items of furniture. There are some interesting points. It seems that people in the lowest income bracket (£3.00 per week or less) would be willing to pay £4.25 for a dining table (more than their weekly wage), while those in the top bracket (£19.00 and upwards) would expect to pay £10.25 (just over half their weekly income). For a sideboard, the £3 per week man would pay £5.50, the £19 per week man £17.25. A dining chair at £0.96 was regarded as reasonable by the lowest paid, but the top group were prepared to pay £2.32½. In general, the higher-paid group were willing to pay roughly twice as much as their colleagues in the lower brackets for each item mentioned. 'Top' prices for a new double bed or divan average at £11.20, for a single bed £7.50, for a settee £13.25 and for a wardrobe, £14.20.

The prices *Frank Murphy Ltd.* proposed to aim at were:—

Dining table	£4.25	—£7.50
Dining chair	£0.99	—£1.75
Sideboard	£5.50	—£9.75
Easy chair	£3.75	—£6.75
Settee	£7.25	—£12.75
Double bed	£5.75	—£10.00
Single bed	£3.37½	—£6.00
Wardrobe	£7.00	—£12.50
Chest of drawers	£4.00	—£7.00
Dressing table	£5.00	—£9.00

(figures decimalised for easier comparison).

These would be within the compass of people earning between £150 and £600 per annum—that is, about 80%—90% of the population at that time.

Under 'General Finance', readers of the *Murphy Review* were given a statement on the share capital and number of shares so far taken up (see page 161), a statement of accounts up to September 3rd 1938, (cash at Bank and in hand £5,314 15s. 9d.), a detailed approximate weekly expenditure, at that date £70.00), income from sale of products (Nil), and a schedule of salary grades. The lowest grade out of the ten proposed had a salary range from £1 to £1 5s. 0d. per week and the top grade from £23 10s. 0d. to £28 10s. 0d. per week.

Names of the staff already appointed, their salaries and their designation were then given:—

Grade	Salary	Name	Duties
A	£1 5s. 0d.	D. M. Hulks	General office
A	£1 5s. 0d.	S. M. L. McGeehan	General office
D	£3 10s. 0d.	C. J. Chiverton	General office
E	£4 10s. 0d.	J. G. Naz	Materials
F	£6 10s. 0d.	A. Rowcliffe	Personal secretary to Governing Director

F	£5 15s. 0d.	A. G. Hood	Mechanical transport
G	£8 5s. 0d.	K. J. D. Murphy	Market Research and retailing
G	£8 5s. 0d.	E. Holliday	Company Secretary, Accounts, General office
G	£8 5s. 0d.	J. D. A. Boyd	Materials
	Nil	Frank Murphy	Governing Director

Another point revealed in the section on Finance was that Frank Murphy had paid out some £4,000 of his own money in the preliminary expenses of setting up the Company, and it had been agreed that while in the initial stages of its development he would draw no salary, the repayment of these expenses would be made by instalments, with an initial payment of £300, and thereafter payments of £40 per week.

The letter headings of the new Company gave as its headquarters; Ludwick Corner, Hatfield Hyde, Welwyn Garden City, Hertfordshire. This was, in fact, the Murphy's family home, to which they had moved in the expansionist Murphy Radio days of 1936. Although Ludwick Corner was within the Welwyn Garden City postal area, it was then isolated some distance eastwards from the main part of the town and as the nearest houses were in the council housing estate of Pear Tree, Frank Murphy used to remark with a grin, "I've decided to cross the railway to join my workers!"

Actually, the house he and Hilda chose, though not old, was of some architectural and historic significance. The house dated from about 1910, and the family were always told that Lord Salisbury had caused it to be designed by an architect called Newton who followed the tradition of Lutyens. The purpose of the house, so it was said, was to accommodate a certain Mrs. Tweedie and her children, in whom it was alleged Lord Salisbury had more than a neighbourly interest. However that may be, there certainly was a scratched name "Tweedie" on one of the bedroom windows, and when the floor-boards creaked on the upper landing, we always joked that the ghost of Mrs. Tweedie was walking there.

Ludwick Corner's distinctive features were two rounded porches at the front entrance and the door to the patio and back garden. Twin gables marked the east and west wings of the house, while the eastern side was extended to provide living accommodation for the staff, behind which there was a yard and ample stable or garage accommodation (see Fig 13). Most of the rooms were at least 20 foot square (much bigger than conventional Garden City homes), and for children there was the excitement of racing up and down long wide corridors, and exploring out of doors in the extensive gardens and adjacent spinney.

Unusually for those days, the house had full central heating (perhaps Mrs. Tweedie was an American), and the lounge also had a large inglenook fireplace where a log fire was maintained on wood ash. This was the favourite family room for charades and games at Christmas time, because one could sweep aside the heavy velvet curtains dividing it from the corridor, to make an imposing stage entrance.

Beyond this there was a sitting room with one wall completely lined with book shelves. This housed Frank Murphy's prized and much-used possession—a complete edition of the Oxford English Dictionary in some forty-four volumes. At first he used the room as a billiards room, with a three-quarter size billiards table taking up most of the space, but later the table was removed, and the room became a comfortable sitting-room-cum-study, where many of Frank Murphy's friends will remember hours of enthusiastic talk and argument.

The drawing room was a large pleasant airy room overlooking the garden, and I think it was one of Hilda's biggest personal sacrifices to give this up to Frank Murphy Limited, and to see it converted into a workshop and designer's laboratory. The Gordon Russell furniture was moved out, the Wilton carpet and the curtains removed, and carpenters' benches and tools moved in. (See fig. 36).

The open fire-place room suffered in a similar, but not quite so drastic way. The easy chairs were moved out, to make way for experiments testing the strength of the legs of the prototype dining chair, and the dining table's surface resistance to scratches (see fig.45).

Upstairs, two small spare bedrooms were taken over to serve as offices, but otherwise there was no very visible change. There was plenty of room for the family to live in the eastern end of the house, especially as Maurice and Joan were away at boarding school and college respectively. In the dining room which, like the former drawing room, overlooked the garden, the visitor's eyes went immediately to the beautiful circular dining table of walnut and bird's eye maple, designed by Gordon Russell. With its central pedestal, it was inevitably known as 'the mushroom'. There was a matching Russell sideboard and a serving table, and an elegant plain green carpet, with a Marian Pepler rug.

As one looked out at the two lawns (one large enough to hold a tennis court) and at Hilda's colourful herbaceous borders leading down to the little pond and the 'wild garden', with the big silver birch tossing its leaves against the window, it was easy to see why Ludwick Corner was such a delightful home and work-place, and why in December 1938 the Company applied for planning permission to develop 200 acres adjacent to Ludwick Corner as a landscaped research centre.

This request was refused by Welwyn Garden City Urban District Council, on the grounds that Ludwick Corner was in a residential area (undeveloped, it is true, at that time) and that such activities could be more suitably carried on in an industrial area. At the subsequent Public Enquiry, Frank Murphy's case (supported by photographs) was that even if he was interested in building a factory (which he was not), he would not be prepared to set it up in the Garden City factory area because that was too ugly. "The staff were asked if they would prefer to work looking out on trees or on bricks and mortar, and they answered that they preferred trees. They also agreed that they would turn out better work in these surroundings . . . Manufacturers like myself are all over the country trying to persuade the working class to buy beautiful objects, but without much success. If as a country we immerse our workers in ugly surroundings for five days a week, nature will give them a protective remedy and make them insensible to beauty." (*Murphy Review* No. 7, p. 338).

Needless to say, the Ministry of Health upheld the decision of the Urban District Council to refuse the Company's application and Frank Murphy lost his appeal. He had committed the unforgiveable sin of accusing the Garden City planners, of all people, of creating an ugly environment, when for years they had considered themselves, in many ways rightly, the pioneers of enlightened urban design. That they had given the workers' housing on the eastern side, a far less pleasant environment than the middle-class commuters on the west side had somehow escaped their notice, and they were not pleased to be told so publicly.

After the war when the New Towns Act became law, the old Garden City Company was superseded by the Welwyn Garden City Development Corporation, which was charged with the task of building homes for hundreds of Londoners and ex-servicemen. So the area around Ludwick Corner was scheduled for development, and Ludwick Corner itself was converted into flats. The "wild garden" disappeared for ever, but someone with natural taste decreed that the lawns should be saved and converted into an open space like a village green, with new detached houses grouped round it. The area was called Beehive Green, after the old public house, the Beehive, at Hatfield Hyde, and today Ludwick Corner, externally at least, looks much as it always did. The birch trees and the twenty-foot high magnolia by the patio have survived, and I hope give as much pleasure now as before. (See fig. 14.)

By mid-October 1938 three new members of staff had joined the Company; A. J. Sims, the cabinet-maker who, with W. C. Wood (who joined the following January), was responsible for making the first tables, chairs, and sideboards to designs worked out by J. D. A. Boyd. Another new member was Rosamund Willis, a textile designer whose occupation was given rather curiously, as "Aesthetics". The third newcomer was Eric R. Thomas, formerly in charge of Dealer Publicity at Murphy Radio, who was now to become responsible for printing, advertising and editing the *Murphy Review*.

The second issue of the *Murphy Review* was published barely a month after Chamberlain's dash to see Hitler at Munich, so it was natural that the first article should be headed "War, Peace and Neutrality". In it Frank Murphy sought to define the three words in a way that would help a person to see how he could do his individual bit to seek peace; and incidentally escape the helpless feeling of being dragged into war by events over which he had no control.

Characteristically, Frank Murphy asserted that "nations, no more than buildings, cannot be at war, but persons individually or collectively can be at war with other individuals or with each and every individual forming another group or nation.". . . "I can only express good or bad will to my neighbours by performing towards them economic acts... To be at peace with the world (i.e. the sum total of the individuals living in it), I must at all times serve the consumer. I can do so only in so far as I acquire more knowledge and apply it honestly to his problems, or, in the words of the New Conception of Business, I must apply knowledge with integrity in the service of society".

"Therefore the peace of the world is not a national or state problem, but a problem for each and every individual. If in our daily lives we offer knowledge applied with integrity to every person we contact or, alternatively stated, seek his or her advantage, although we cannot guarantee such a course will ensure the peace, it has a better chance than any other of so doing."

After this rather tough beginning, the readers were treated to a little light relief in the shape of a joke.

>FISHING AT GODESBERG
>
>The following story comes from Germany:—
>
>The negotiations between Hitler and Chamberlain looked like coming to a standstill, so they decided to have a little relaxation in the way of fishing.
>
>They proceeded to the Rhine, but whereas Hitler had no luck at all, Chamberlain kept pulling out fish after fish.
>
>Hitler naturally was a bit upset and asked Chamberlain how he managed to do it. 'Ah', said Chamberlain, 'I get a lot of training, because you see in my country we let the fish open their mouths!'"

Strangely enough, this harmless story upset several readers, who wrote in to complain that it was "in rather poor taste". However, to me it seems to be in much the same vein as the famous wartime song "We're going to hang out our washing on the Siegfried Line, if the Siegfried Line's still there!" and any offence lies in the blandly unconscious assumption of superiority by the British.

The Table of Contents for issue 2 makes rather quaint reading, with its extraordinary mixture of topics:—

>The Honorary Board of Trustees (Names and addresses)
>Editorial Note (to encourage comments from readers)
>War, Peace and Neutrality (Frank Murphy)
>The New Conception of Business—How it works (one of the staff)
>Memorandum to the staff (Frank Murphy)

Fishing at Godesberg (See above)
This Freedom (comment on transfer of labour without the individual's consent, in a company take-over)
Table Design (J. D. A. Boyd)
Good Beds (Frank Murphy)
Aesthetics (Frank Murphy)
Retailing (Kenneth Murphy)
Statistics
Statutory meeting
General Finance
Correspondence (between Trustee A. R. Turner and Frank Murphy)
Book Notices ("Civilisation" by Clive Bell, reviewed by Frank Murphy).

At first sight, it looks rather a one-man show (no doubt unavoidable when the Editor-to-be had not yet taken up his duties) and overweighted with philosophic speculation. It should however be pointed out that the thoughts on freedom, war and peace occupy only six pages, while the "practical" articles, beginning with Table Design, take up twenty-five pages, and the correspondence and Book Notice four pages.

"Table Design" was the first of Douglas Boyd's articles to appear, in which he at once announced that he would consider the subject from the engineer's point of view as distinct from the aesthetic. He wrote in a simple and readable way about the choice of materials for tables and then dealt with their construction, deftly bringing in both history and humour, as the following extract shows:—

"Early English dining tables (fifteenth century) were long planks of wood lying loosely on several trestles, which could be packed away when not in use. People nowadays, however, seldom have large banqueting halls, and where a few odd trestles were neither here nor there beneath a thirty foot plank, they would be clumsy in a small room. Nowadays it requires a more knowledgeable use of supports to save cost and weight and give strength economically."

He goes on to illustrate graphically the practical advantages of placing the legs of a rectangular table either at the corners, or set in, as in a refectory type table. Two photographs accompany the article, one of an early seventeenth century drawleaf table in the

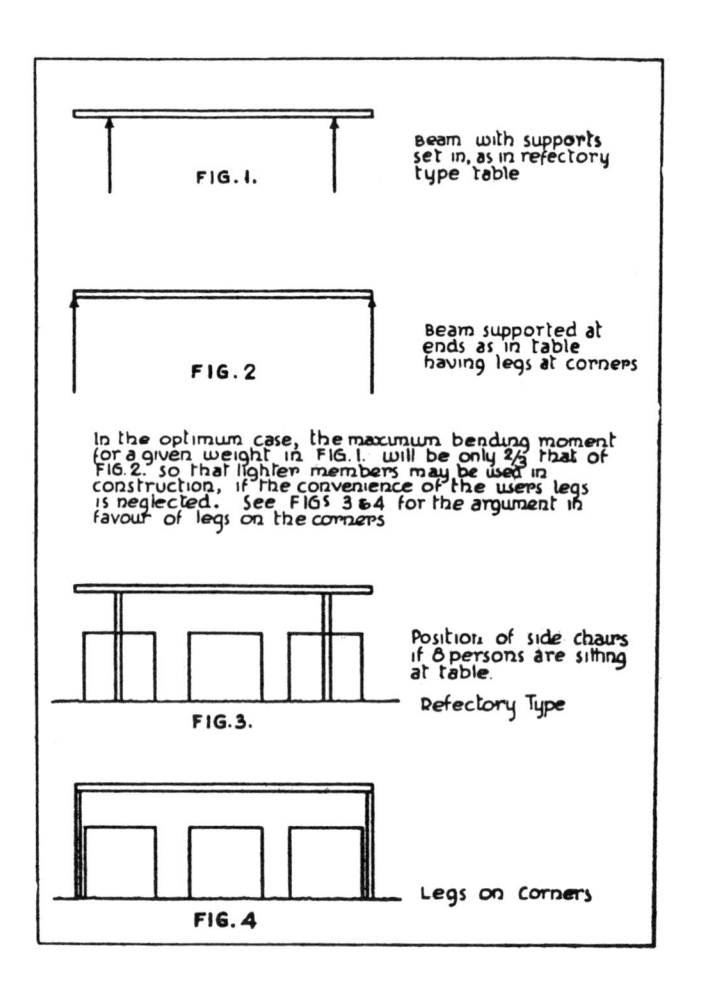

FIG. 1. Beam with supports set in, as in refectory type table

FIG. 2 Beam supported at ends as in table having legs at corners

In the optimum case, the maximum bending moment for a given weight in FIG. 1. will be only 2/3 that of FIG. 2. so that lighter members may be used in construction, if the convenience of the users legs is neglected. See FIGS 3 & 4 for the argument in favour of legs on the corners

FIG. 3. Position of side chairs if 8 persons are sitting at table. Refectory Type

FIG. 4 Legs on Corners

Victoria and Albert Museum, and the other of the first refectory table produced by Frank Murphy Limited.

Pointing out that as well as making the table structure strong, reliable and convenient to sit at, it was also important to protect the surface of the table from domestic spills and accidents, Boyd disclosed that the Company had found a protective polish which withstood the following tests:—

1. Boiling water poured on the table.
2. Whisky, tea, port, beer and ink upset on it.
3. Ironing a wet tea-towel directly on it.
4. Standing a metal teapot full of hot tea, the electric kettle with boiling water; and a Pyrex dish straight from the oven on it.

This was no idle boast. John Pank—a former Murphy dealer-recalls the meeting in Norwich when Frank Murphy gave a practical demonstration of the properties of the polish, by boiling water in an electric kettle and pouring it straight on to the table, spilling ink on it, and ironing a wet teacloth on it—all with no ill effects. This was long before the days of universal polyurethane protection for furniture.

Boyd's article on tables was followed by Frank Murphy's equally practical article on "Good Beds".

He began with a typically disarming sentence. "The average person, if he is anything like me, must find it very confusing when he tries to buy a bed as the catalogues don't tell you how much a bed costs. They tell you the price of over-lay mattresses, box mattresses and divans, but never the price of a bed."

He then went on to detail the raw materials used in making beds (steel, rubber, fibres, feathers, etc.) and summed up his problem: to find out which material gave the best bed for a given sum of money.

But what *was* the 'best bed'? Once more, the engineer's approach produced some surprising answers. Having first noted the lack of unanimity of opinion, either among the public or the furniture trade, as to whether steel spring mattresses or rubber mattresses were superior, he asked himself the usual question, "What is a bed for?" and began to compare beds with easy chairs. In the case of the latter, he noticed that the most comfortable ones always had a down cushion resting on a steel cushion—that is, both materials were used. This led him to the conclusion that the problem in making a good bed was not essentially a problem in springing, but of providing a surface which would adjust to the shape of the body and give the maximum bearing surface to carry the weight.

"I suggest that we measure comfort, or rather discomfort, by the pounds pressure per square inch on the body, and because this is very high when sitting on a gate, we say that it is uncomfortable. Alternatively, the most comfortable bed is that provided by the sea when floating on your back in it, because then the weight of a body is distributed over a very large area and the pounds pressure per square inch is very low."

After this basic requirement of adjustability, Frank Murphy then listed four other requirements of good beds:—

"2. They must provide the requisite heat insulation under the body.

If too good a heat insulator is provided, the sleepers will complain of excessive warmth, unless they normally unduly feel the cold, when I suggest they will want the heat insulation provided by a feather mattress. If too little heat is provided, they will feel cold all night, despite the number of blankets on top of them. Some of us have had this experience when sleeping in camp beds with no blankets under us. (A reminiscence of Royal Flying Corps days, perhaps).

3. They must be elastic, but not perfectly elastic . . . The material must show what is called 'hysteresis', that is although it should go back to normal when the pressure is removed, there should be a time lag between the return movement and the removal of the pressure (e.g. as in the case of a down cushion).

4. The materials should be non-inflammable, as some people smoke in bed.

5. The materials should continue to function properly for 10-20 years."

On adjustability, he pointed out that at that time rubber cost 7d. a pound and steel only 1½d. to 2d. a pound, so steel springs would achieve the same result more cheaply than rubber.

Steel gave little heat insulation, and rubber latex too much. But you could cover steel with a non-conducting material such as wool or some other fibre which would make it acceptable for heat insulation.

Similarly, neither steel nor rubber would meet the requirement of imperfect elasticity, but animal fibres such as hair, wool, feathers or down would.

Steel was certainly a non-inflammable material, and fibres were reasonably so, but rubber latex was definitely out on this score. (We can be certain that no Murphy settees, if they had ever been made, would have been upholstered in inflammable materials, as, sadly, so many still are today.)

All the materials,if properly handled and fabricated would, in Frank Murphy's estimation, meet his requirement of durability.

So on the whole he came to the conclusion that some combination of steel and fibres would prove the most suitable raw material for a good bed.

Two more pages followed, with diagrams, on the different methods of springing, horizontal and vertical; on parallel-sided and hour-glass shaped springs; and on the variety of types of steel wire used for springs. After which, the article finished quite characteristically:

"There are probably a dozen ways of making a comfortable bed; the hard problem is how to make a reliable bed at a reasonable cost, i.e. the bed that stays comfortable; and this statement I feel makes a reasonable conclusion to this article which is already long enough."

FRANK MURPHY

A few weeks after J. G. Naz joined Frank Murphy Limited, he was given the job of investigating materials, so it was he who wrote the later articles on steel springs for beds. These are obviously the result of much detailed and conscientious research and contain plenty of good material, but they lack Frank Murphy's originality and home-spun humour and his ability to appeal to the reader's own experience.

The quality of being able to laugh at oneself is not always given to Chairmen and Managing Directors, so it is refreshing to find in a later issue of the *Murphy Review* a short item entitled *A Reply to the Bed Questionnaire:—*

"Mr. Frank Murphy, who with admirable caution always addresses me as Dear Sir or Madam, has suddenly thrown a series of pertinent questions at me.

These are rolled into one disconcerting salvo under the title

of Bed Questionnaire.
Says Mr. Murphy rather abruptly—
Do you like to read in bed?
Me: Yes, please.
Mr. Murphy: Do you eat or like to eat in bed?
Me: I wouldn't be answering this if I didn't eat, and I won't knock you down if you bring up the b. and e. at breakfast time for I have a distinct penchant for egg on the blanket.
Mr. Murphy: Do you like to be able to sit on the edge of the bed whilst undressing?
Me: I can usually stand after closing time, thank you.
Mr. Murphy: Do you want to be able to store things underneath the bed?
Me: Aha! you old wag!
Mr. Murphy: Do you want to use the bed as a couch in the daytime concealing the fact that it is a bed?
Me: I wish to lie in bed all day, and you can call it a bed, couch, divan, or just haybox, so long as you don't wake me.
Mr. Murphy: If you are married, would you choose a double bed or two single beds?
Me: Is nothing sacred to you, sir?"
"Cassandra"
(Reprinted from *The Daily Mirror*, Dec. 22nd 1938).

This last jibe was a true word spoken in jest, for, as has been shown, Frank Murphy loved nothing better than taking devastating pot shots at sacred cows.

However, he was not alone in this. His family, too, though united in their intense pride in his achievements, were quick to spot the weaknesses and foibles of this extraordinary man, and never failed gleefully to point them out if they could, hoping that this would counteract the atmosphere of uncritical adulation which tended to surround him.

Thus, in the very first issue of the *Murphy Review* there appeared a short poem written by Maurice, then at the Downs Preparatory School at Colwall, near Malvern, which scored a bulls-eye:—

"Father's Job.
My Father's job
Is to get a job
For all the souls at sea.
They're big souls and little souls,
And there's souls that ought not to be.

And if you say,
"Where's Pa today?"
You can be sure he's gone to see
Some old, old gentleman
Who charges a £3 fee.

Today he's writing
A book of booklets,
On how to run a firm,
But I bet you anything you like, I do
It won't be finished this term.
       Anon (12)

Similarly Joan, who was at this time studying for an English degree at Cambridge, came up with a satire in heroic complets, closely modelled on her favourite eighteenth century poet and satirist, Alexander Pope:—
    (*Murphy Review*, p. 234-235 Vol. 1, no. 5).

"The Heel of Achilles"
O gentles, hearken to my curious lay—
(Or those who think they've time and patience may)—
I sing of one whose glorious fame resounds
Beyond the petty Garden City bounds:
Whose name is heard wherever bread is won
On India's shore, or darkest Islington:
This man, I say, was one who erstwhile made
The fortunes of the great tobacco trade—
(His photographs afford sufficient data,
Example of a cheerful fumigator).
He first among the manufacturers' throng,
Achieved an instrument for sound and song
That seemed to hold together and was strong;
And mounting with his clarion on high,
Began to tell the British Public Why.

Such wild success, such vigorous plaudits came,
That men became aware of Murphy's name . . .
But now the distant heights he seeks to scale
And Furniture and Furnishings assail,
And a vast audience, eager though unseen,
Peruses booklets, puzzles what they mean,
While business men pronounce him gone or green.

Meanwhile the postman, bowed with heavy load,
Steps o'er the threshold of F.M.'s abode,
And piles of letters show increasingly
Delight and pleasure of the great B.P.

But yet, alas, what boots it thus to sing,
Thus to exalt, and make the welkin ring?
For it is true that we must change our note
T' express the warbles of a mournful throat.

We must relate (O tearful task) that now
A secret sorrow sits upon his brow.
A hidden fear torments his noble heart,
And vultures tear his happiness apart.

The reason for this anguish you shall hear,
And while you listen, shed, O shed a tear.
While all these millions thus received his word,
As if it were an oracle they heard,
While thousands daily urged him to proceed,
Praised his attempts, and ratified his creed;
Alone among th' enthusiastic throng,
A single group of people Done Him Wrong.
And these (Alas, awake, Integrity!)
These graceless people were—his Family.
At breakfast, when the morning mail was brought,
They did not silently imbibe his thought,
But sniggering between themselves would say,
"Is it the curtain lecture due to-day?"
Or boldly insolent, would loudly utter
A sharp retort, or murmur, "Pass the butter."
They ridiculed his fancy's dearest child,
Or winked among themselves, looked wise, and smiled.

"A table, this?" they cried, and shrieked with mirth:
"No one would ever pay for what it's worth!"

And unaware of Opportunity,
"What, have it in my dining room?—Not me!"

Fame is no plant that grows on mortal soil,
Said Milton (though he might have saved his toil)—
No Prophet's great, as you may briefly see,
Within the bosom of his family.

JOAN MURPHY.

This produced a reproof in Yorkshire dialect in the next issue:—

"*Ark et Joan*"
Nay, lass, tha' shouldn't laugh
At all thi feyther's gaff,
He means weel as yo' should know,
For all his high falutin' jaw.

Them theer "voices" he ses he hears
Abart table legs an' kitchen cheers,
Are none so gaumless as yo' think,
They've driven many a man to drink.

And some day when he's burnt at stake,
Yo'll say, "Eh! There's some mistake,
It's me as should be theer—not Dad—
For laughing at a gradely lad!"

W.H.B.

Despite these shafts of affectionate criticism, Frank Murphy continued to encourage his staff in their daily application of the ideals behind the New Conception of Business, and in making good quality furniture designed to meet the customer's stated needs.

The front cover of the third issue of the *Murphy Review* (November, 1938) was illustrated with a photograph of the dining table (called a fixed table, as distinct from a draw-leaf or extending table) which the design team had arrived at after much thought and experiment. It had a number of unusual features, all connected with practical utility.

The first requirement was the seating accommodation required. This, according to questionnaire returns, was normally for four people, but on special occasions, for eight. Consequently the table top needed to be 5'6" long by 3'0" wide. This allowed a table setting of 22" per person, if three people were seated on each side and two at each end, and also provided space for the family joint and the vegetable dishes in the centre of the table.

The corners of the table top were rounded to avoid hurting people who accidentally bumped into the table. The legs were placed 3½" in from the edge of the table, so that those sitting at the side in the end position could draw in their chairs without the front edge of their chairs fouling the legs of the table. The height of the table was 2'5", which had proved the most convenient for sitting at with a dining chair with a standard height of 1'6".

All the table legs had the corners radiused—a large radius on the outer corner and smaller ones on the inner corner—on the practical consideration that "if the corners were not taken off, they would very soon in actual use be kicked off".

The rail below the table top was shaped to a greater depth at the ends to ensure a sufficiently strong joint in the leg—the most vulnerable point.

The tenon joint in each leg of the table was divided, so that movement tendancies in the wood might be localised and so that the leg mortice was not subject to an undue area of cut; while the pins through the tenons were staggered to avoid cleavage stress being exerted on the same line from each pin (see fig. 49). The glue used in jointing the quarter sawn table top provided a joint under normal conditions stronger than the wood across the grain, while the top was secured to the trestle with buttons (see fig. 49). Finally two coats of cellulose lacquer were applied over the natural oak top, to provide the remarkable protection already detailed on page 145.

The retail price of the fixed table was to be £5 10s. 0d., and for the draw-leaf table (fig. 43), £6 15s. 0d.

Consumer research on requirements for a sideboard had produced a motley array of articles requiring storage.—Cutlery, tablecloths, table napkins, cruets, wine and beer bottles, sauce bottles, and glassware were universally specified. But the higher income groups apparently expected to house their tea, coffee and dinner services elsewhere, while only the lowest income group seemed to need space for table mats and "sewing".Interestingly enough, when people saw the furniture in actuality at meetings held in February and March 1939 there was general acceptance of the tables and dining chairs (of which more, in a moment), but not of the sideboard as originally proposed. When challenged people said it was too small, yet had to admit that it would hold all they had asked for. Frank Murphy then suggested that what they really wanted was something they had not actually stated, since they were not consciously aware of it. This was, to provide the maximum amount of storage space which their dining room would allow, and that beyond requirements for cupboards and drawers, of which one must be for cutlery, people were not too much concerned with the sideboard.

Eventually, two types of sideboard were produced for sale; a smaller one 4' long and 3' high with two drawers and two cupboards under, standing on a solid plinth. The storage space was given as 21,000 cubic inches, and the price was to be £12 15s. 0d. (see fig. 43).

A larger 5' sideboard standing on legs, with two side cupboards and four centrally placed drawers—the bottom one two inches deeper than the rest—would provide a total storage space of 20,500 cubic inches and would retail at £15 10s. 0d. (see fig. 41).

It is interesting to compare these designs with those for "Utility" furniture produced a little later under Government licence during the war years. The standard of finish and general specification of the Murphy furniture was, however, much higher than could be achieved within wartime restrictions.

Much the most unusual and striking of Douglas Boyd's designs, however, was for the dining chair. Research started with looking at the actual posture and movements of people sitting on dining chairs, and from this observing that the body was held more or less poised on the seat of the chair, and the back only used during pauses. So he deduced that the essential needs were:—

    1. A support for the body at a convenient height for the normal dining table, and a back support for occasional use.

    2. A method of supplying the above with comfort and reliability.

Boyd pointed out that when one is in a sitting position most of the body weight is supported on the bones of the ischium, and as these and the flesh between them occupy a very small area, there is consequently a high pressure in pounds per square inch. Increasing the area of support below the body reduces the pressure per square inch and results in a greater degree of comfort (as

Frank Murphy had pointed out in his article on beds). Now the conventional dining chair had (and still has) a drop-in seat with a frame of either webbing or a sheet of plywood, covered with textile or leather cloth padded with a small quantity of stuffing. A better technical solution, however, seemed to be to provide a concave rather than a convex surface, to spread the support for the pelvis.

Could this be achieved in a hard material, without using textiles or leather for the chair seat? The design team, led by Boyd, suspected that it could. and carried out a series of experiments resulting triumphantly in a remarkably comfortable dining chair made entirely of wood. It is worth while quoting from Boyd's article "Comfort with 'hard' seats" (*Murphy Review* no. 7, March 1939) which not only describes the experiments, but the reasons for conducting them:

> "A tray of modelling clay was prepared and supported at a convenient height, then an impression was taken by having persons actually sit in the damp clay. The impression clearly showed the deep indents produced by the pelvis bones and a slighter one by the base of the spine, and it was very evident that most of the body weight was taken at these points (see fig. 47).
>
> As one does not always sit in one position, and as all people are not alike, another tray ws produced filled with damp sand (see fig. 48); this was hollowed out to a shape based on the original clay seat until it was comfortable in most positions usually assumed on a dining chair. From this, templates were made on a 2″ grid and a seat carved out of a 2″ thickness of ply.
>
> This seat will eventually be pressed in a thin light ply to this shape.
>
> The shape of the chair back was obtained in like manner using a tray of damp garden clay supported on a temporary "lash-up" of a chair.
>
> The overal angle of the back was averaged by trying it on a number of people, before the clay was put in place. Here the clay gave the information that the back should be curved both horizontally and vertically: and this has been done in the sample illustrated.

You knew me as Frank Murphy of Murphy Radio Ltd, but it is over two years now since I was with that company.

I have happy memories of you. For instance, one day in an Essex town, a window cleaner came off his ladder and said, "Excuse me, but aren't you Mr Murphy?" I said, "Yes"; to which he replied, "I'd like to shake hands with you; I have got one of your sets."

That handshake gave me a lot of pleasure, and now that I am offering you furniture instead of radio, I hope you will give me the same friendship.

Yours sincerely,

Frank Murphy.

Fig 31. Inside Cover of Frank Murphy Ltd. Furniture catalogue

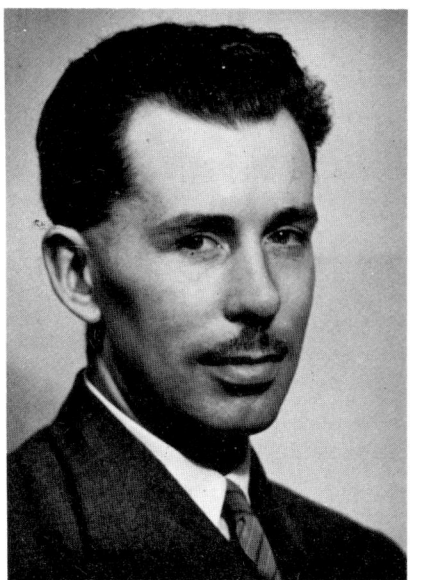

Fig 32. Kenneth Murphy (1938)

Fig 33. Audrey Murphy (1932)

Fig 34. The staff of Frank Murphy Ltd at Ludwick Corner (1939)

Fig 35. A. R. Turner (1938)

Fig 36. The Cabinet-making shop at Ludwick Corner

Fig 37. Old Windsor Chair

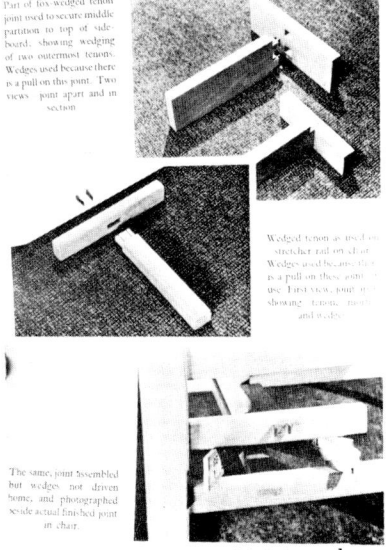

Fig 38. Types of joint used

Fig 39. The Murphy dining chair

Fig 40. The fixed dining table

Fig 41. The 5-foot sideboard (1)

Fig 42. The 5-foot sideboard (2)

Fig 43. The 4-foot sideboard, drawleaf table and dining chairs

Fig 44. Test for scratching of table top surface

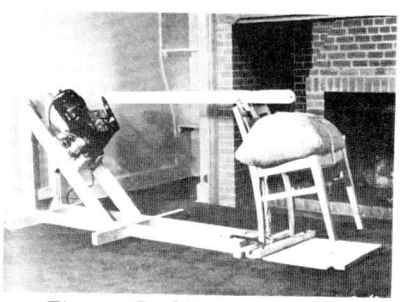

Fig 45. Racking test on chair

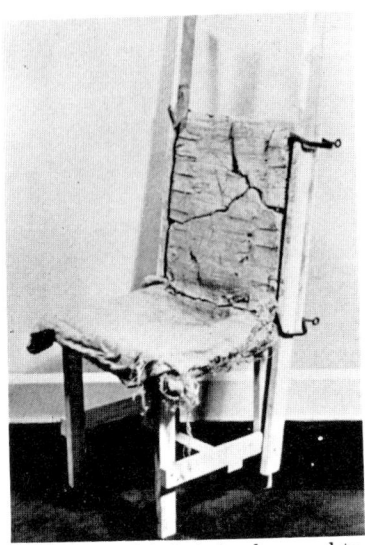

Fig 46. Model chair in clay used to find correct shape and angle for back

Fig 47. Mould in modelling clay to ascertain correct seat shape when bodily position is unchanged

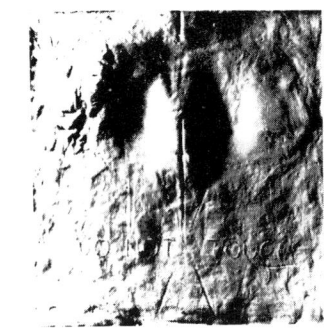

Fig 48. Mould in sand based on previous clay mould, but allowing for change in bodily position

Fig 49. Underneath of fixed table

Fig 50. Mechanism of drawleaf table

The height of the back has been kept below the shoulder blades, which gives ample support without discomfort and allows turning and stretching backwards over the chair for an object, which would not be possible with a really high back. The latter would add to the weight and cost of the chair without a commensurate increase in comfort.

It will be noted that the back legs are left a maximum thickness at the joints with the seat frame, and are reduced in thickness above that point on the portion framing the back, in order to save weight, and this has been done on the *outside* to allow as much freedom for the elbows as possible. On the other hand, below the seat frame the reduction has been done on the *inside* in order to keep the leg base as wide as possible for stability reasons.

As the user often tilts the chair back, or drops the chair on its front legs, stretcher rails are used to increase the strength of both front and back legs relative to the seat frame. They are kept as high as possible consistent with strength considerations, in order to facilitate sweeping the floor around the legs.

It is interesting to note that our forefathers are reported to have got the shape of the Old Windsor chair seats by methods similar to our own; and I have it from an old London chair firm, that the Old Windsor seats were anything up to 1½" deep (actually our own independent result is 1½") and we ourselves have since measured Old Windsor seats of 1 and one sixteenth of an inch deep (see fig. 37).

The modern so called Windsor chair, with its purely nominal hollowing of the seat, is a travesty of its forebears.

The Murphy chair described is very definitely comfortable and supports the body in an effortless upright position, without the 'slummocking' which occurs with the usual pincushion seat."

J. D. A. BOYD

By May 1939 the Editor of the *Review* was able to tell shareholder and prospective customers that the fixed and drawleaf tables were now in production, that efforts were being made to find a manufacturer capable of mass-producing the dining chairs to the Murphy specification, and the small sideboard was being re-designed as a result of the meetings earlier in the year which showed that the first version was not what the consumer wanted —it was now at the sample stage, as was the larger sideboard, and both would be referred back to the consumer for approval, or if necessary modification, before proceeding to production stage. All the dining room furniture would be available in light (natural oak), medium and dark shades. The medium colour woods eventually chosen were Australian walnut and Guarea, or Nigerian pearwood, which had a warm red tone reminiscent of unstained mahogany. These satisfied the design team's stringent requirements e.g. availability quarter-sawn in suitable widths and lengths, and freedom from defects and waste. The movement (shrinkage and expansion) had to be similar to that of prime American white oak. The wood had to be hard, that is resistant to bruising and marking; to glue satisfactorily; and had to take the finishing lacquer satisfactorily.

By October 1939 200 tables had been made by McIntosh of Kirkcaldy and were ready to be distributed to warehouse centres; 200 sideboards were being made by the Educational Supply Association Ltd., at Stevenage, with completion promised in a further six weeks.

A manufacturer had been found for the Murphy dining chairs and after some delay an order was placed for 350.

Meanwhile further questionnaires and interviews had established the consumer's needs in the way of bedroom furniture. A sample light oak dressing table with a full length mirror had been made and photographed and readers were asked to give their reactions to it. Sample wardrobes were also prepared, and of three suggested arrangements for a chest of drawers most people had shown a preference for a unit 3'6" tall, 2'6" wide containing three drawers above which would be a 1' deep cupboard, with opening doors.

The art of framing questions for a questionnaire is not simple as Douglas Boyd ruefully admitted in June 1939. He retailed how the design team had asked what they thought were straightforward questions on what people wanted to keep in a sideboard. The answers were also straightforward. The trouble was, that people put down all the articles they thought ideally they *ought* to keep in their sideboard, not what they actually did keep in it. So when the designers produced the first sample sideboard which met all the stated requirements, they were surprised when people said they didn't like it. Asked why not, the answer was a rather incoherent "We don't like the looks of it".

Actually, Boyd detected the real reason for people's instinctive rejection of the sample sideboard.

"It was this—you wouldn't have enough top room—or 'sideboard' and not enough room for storing those old gramophone records, the odd empty case of fish-carvers that you 'just couldn't throw away', the flex for the percolator, the odd writing materials and the half stone of dog biscuits for Fido!'

You don't keep such things in your sideboard?—Oh yes you do, we can prove it!

Of course, you would never put a tray on your sideboard!- But you do really, when the table is half laid and you can't put the tray on that dinner wagon which you mean to have some day, but actually haven't got yet!

Honestly, though, this referring back to you has been most illuminating, and we now know not only more about the *real* requirements of a sideboard, but also more about you, Mr. and Mrs Consumer. This will help towards an even better fulfilment of your needs in other things.

In the long run, by first getting a reasonably sound knowledge of your needs and then fulfilling this within your salary scale, you will get *a higher value for money ratio* (note the old familiar phrase) uncomplicated by the losses incurred in splitting a given quantity into fifty different shapes and sizes."

Chapter Eleven

MARKETING PLANS FOR MURPHY FURNITURE

"The best-laid plans of mice and men   Gang aft agley . . ."
                                                                                                  Robert Burns.

July 1939 saw the publication of the first illustrated catalogue of the furniture. Typically, it did not merely contain photographs of the tables, chairs and sideboards and their respective prices, but a brief introduction by Frank Murphy (see fig. 31), a page explaining the Company's primary objective "to apply knowledge with integrity and so express one's individuality in the service of society . . . We shall therefore only make such furniture as enables this to be done".

Then, and only then, came the pictures and the descriptions of the furniture, and flowing naturally out of the decriptions came explanations of the reasons behind the various points of design, photographs of the *underneath* of the tables to show construction details, photographs of the experiments leading to the design of the chair, and photographs of tests on the security of the chair's back legs and on the resistance to scratching of the table top (see figs.   38-50).

Two other items were included in the catalogue which were unique. One was the statement that the prices quoted for the furniture allowed for a 'plus tolerance' or margin, to cover the costs of retailing which could not then be accurately known. If it turned out in practice that the prices quoted more than covered the costs of making and retailing, any surplus would be duly repaid to purchasers of the furniture—or retained to their credit for future furniture purchases or the purchase of Company shares if so desired.

The second item concerned the Company showrooms. Explaining that the true function of a shop was not to "sell" but to show goods, and allow people to see and handle what they wanted to

buy, the public were informed that Murphy showroom staff would be paid fixed salaries not paid on commission and they would be paid fixed salaries on commission and they would have three duties:—
- (1) to give the public such information as they needed about the furniture.
- (2) to ascertain their needs in this respect ("They will not tell you what you ought to have or ought not to have. Their job is to help you to get what you desire.").
- (3) to give information about the New Conception of Business.

whatever was ordered would come straight from the factory, and every endeavour would be made to keep production as close as possible to sample.

Meanwhile the retailing side had not been neglected. As has already been noted in the furniture catalogue, Frank Murphy Limited had decided to show samples of their furniture in major centres throughout the United Kingdom and to make deliveries of orders from stocks of manufactured items in one or two larger warehouses. Believing that the consumer did not wish to be 'sold' anything—certainly not in the sense of being persuaded to buy against his better judgment, the Company offered to Murphy dealers and other suitable staff the posts of showroom managers and assistants. Their task was to be informative both about the furniture and about the N.C.B., but equally important, they would be 'listening posts' to convey back to the technical design team any faults found, or needs which had not been satisfied.

This scheme was arrived at after nine months of discussion with numerous people with experience of retailing. Although Frank Murphy provided the initial stimulus, credit for the detailed slogging away at practical problems, and the sympathetic handling of keen but apprehensive dealers must go to his son Kenneth.

As early as September 1938 he wrote to a number of furniture retailers inviting their comments on a proposed scheme for marketing Murphy furniture. This involved the Company renting space from the retailer, to include any to him of lighting, heating, cleaning and rates. The retailer, also proposed to hire a number of assistants from him in addition the furniture, take the customer's money, the Company were prepared to pay 10% to cover any miscellaneous items and pay him a return on his capital involved.

These proposals had a number of difficulties from the retailer's point of view, some of which could be overcome (e.g. difficulty in separating Frank Murphy Limited furniture, from other furniture stocked): but the major stumbling block which could not be overcome, was the Company's intention of working a system of consumer dividends (see page 74). Clearly, it would be invidious for such dividends to be paid in respect of some goods sold in the shop, but not on others.

It thus became evident that Frank Murphy Limited, would have to open their own showrooms and operate them with their own staff. This obviously called for a considerable influx of new capital, not very easily attracted at this early stage of the Company's existence.

Consequently a number of Murphy Radio dealers who had shown serious interest in the ideals of the New Conception of Business were invited to attend an all-day conference at Ludwick Corner on April 24th 1939, to consider how they could join Frank Murphy Limited. Four former representatives had already joined —T. L. Owen, J. Patterson, J. B. Ridley and J. C. Robertson—and they, as Regional Managers, would visit all the 1200 existing Murphy Radio dealers and ask for support, either by providing capital, or by joining the staff themselves.

The report of the conference (*Murphy Review*, May 1939) written by Frank Murphy in his usual workmanlike style lists the options open to those wishing to join the Company. They could

(1) show the furniture in their present shops as Frank Murphy Limited shops;

(2) man the showrooms which it was proposed to open in London and all the big regional centres, closing down their present shops entirely;

(3) close down their present shops immediately and come forward to help raise new capital.

The meeting agreed that it

(1) was not practicable since most Murphy dealers' shops would not be located in the right position in a regional city centre;

(2) was therefore the ... necessity for disposal solution, but the timing (and the ... the dealer's own radio shop)

depended on individual dealers' circumstances.

Frank Murphy then stated:

"As regards raising capital, the evidence to date is that one in two of the Murphy dealers is willing to put up capital, provided he is called on by a person in whom he has confidence. The average amount of capital is £25 per dealer. There are approximately 1200 Murphy dealers, and between them they should be good for 600×25, say £15,000 . . . Dealers willing to call on other dealers would, I believe, meet with a favourable reception. At the same time, they would be able to ascertain which of the 1200 dealers really wanted to join the New Conception of Business, and of those, which were really desirable".

It is not difficult to imagine the alarm and consternation, and even fury, which must have been felt by the Directors of Murphy Radio, when news of these latest proposals of Frank Murphy filtered through to them. No sooner had the wretched man severed his connection with his old firm, which was now firmly re-committed to its primary task of making and selling radio sets, than he was about to re-appear on the scene with an audacious 'take-over' bid for many of their best dealers. All those musings on the individual's rights and liberties, which had seemed to send him safely into orbit in the philosophical stratosphere, were suddenly parachuting him back into their midst. They certainly couldn't afford to let him play ducks and drakes with Murphy Radio again. No doubt Ludwick Corner was not the only place in Welwyn Garden City at that time where conferences were being held!

A month later, on May 22nd, another meeting took place at the Frank Murphy Limited headquarters, when the following practical decisions were made:—

(1) *Advertising.* There had been two likenesses of Frank Murphy used when he was head of Murphy Radio; one, a black and white silhouette of his profile, the other, various photographs of his head and shoulders. It was agreed that it would be undesirable and unwise to use the black and white silhouette as this fell into the category of a trade mark or symbol, which might in time be regarded by the public as more important than the New Conception of Business. A simple natural photograph could,

however, be used, though this too could be discontinued as time passed.

(2) *Sources of further capital.*
(a) Murphy dealers who were already shareholders, subscribers to the *Murphy Review*, or neither of these things
(b) Other shareholders and subscribers
(c) Customers of those dealers who were shareholders
(d) Owners of Murphy sets
(e) Other members of the general public.

(3) *Murphy Review.* In response to many subscribers urging wider publication it was agreed that all present should urge their local newsagents to stock the magazine and put it on public display.

(4) *Furniture.* Douglas Boyd reviewed the current designs for sideboards and tables in accordance with general preferences which had been previously expressed.

By June 27th, Frank Murphy had sent out a circular letter to all those concerned, telling them that in his view the next step was not, as previously thought, to find the capital to sell the furniture; it was *to build the selling organisation to sell the furniture.* This would, of course, cost money, but not nearly as much as the capital required to do the actual selling, and once the organisation was in being, it would breed the finance to sell the furniture.

He analysed the growth of the Copmpany so far as falling into three phases:—

(1) From January 1937 to June 1938: hammering out the *Agreement* which made the N.C.B. a practical possibility.

(2) From June 1938 to June 1939: the production of (a) dining room furniture; (b) the *Murphy Review.*

(3) From June 1939 onwards: the creation of the *selling organisation* by means of a voluntary merger of all those Murphy dealers who in the old days would have been described as "Murphy Mad" and any other persons who wished to sell a worth-while product under the N.C.B.

Practical steps towards the implementation of Phase Three were discussed ten days later at Ludwick Corner, at which some optimistic statements were made. First, those dealers ready to join the merger, having signed the Agreement to operate under the N.C.B., should then take steps to dispose of their businesses, making weekly contributions to Frank Murphy Limited, as their capital became available. It was thought that 100 dealers could be counted on within the next four or five weeks with an average capital ultimately available of £200 to £400 each. Since only dining room furniture was as yet ready for sale, the economic number of showrooms to cover England, Scotland, and Wales was probably somewhere between 50 and 100.

The next meeting was scheduled for Monday July 10th and thereafter weekly meetings on Mondays would be held for the benefit of dealers who had not yet decided to come in but who wanted to enquire further into the matter.

At the end of July 1939, the section on General Finance revealed that new share capital was coming in only at the average rate of £206 per week, and sales of furniture and the *Murphy Review* were negligible as yet. Ominously, weekly expenses amounted to £334 per week, and there were unpaid invoices amounting to £873, staff salaries due but not drawn amounting to £666, and the cash at Bank and in hand amounted to only twelve shillings (60p).

However, as the first of the dealers began to merge their businesses and stock with Frank Murphy Limited, things looked considerably healthier, and by August 29th, 1939, the firm's accountant T. H. O'Brien declared that his composite balance sheet showed a surplus of £7,419 of net assets over net liabilities. He estimated that the firm could anticipate an annual turnover of £44,000, since they now had 13 showrooms and 9 new local managers. But fate—and Hitler—decreed otherwise.

Four days after O'Brien's optimistic forecast, England was at war with Germany, and the future of all companies not directly connected with the war effort was in peril. Domestic furniture and radio sets could not be so regarded, which was bad enough; but probably the most damaging effect of the outbreak of war was the severe loss of confidence in the business world and the consequent drying up of sources of credit to a newly-developing firm.

Nevertheless, Frank Murphy wrote, in the September 1939 issue of the *Murphy Review*:—

"Whether one is in favour of the War or against it, I think it is all important that *business should be as nearly usual as possible.*"

Since the War had to be paid for, he assumed that everyone not in the services would be required to work (i.e. there would be little or no unemployment); and that probably up to 25% overtime would be the norm. "The general picture I have in mind is that we shall all be working much harder and obtaining a slightly lower standard of living than is normal in peacetime."

He guessed that the Government would not allow radio sets to be made after Christmas 1939, since they would consider there were enough radio sets already to enable everyone to listen to the news. So the local managers were told to enlarge and concentrate on the service side of their radio businesses. As for furniture, he expected that there would be difficulty with timber suppliers over delivery dates, price increases, and inability to meet the Company's specifications since the best timber would be reserved for war needs. However, Frank Murphy Limited already had 200 tables, and 200 sideboards nearly ready, and they hoped soon to have 350 chairs. So production might be envisaged at least on a modest scale. But first, the dealers had to collect the cash due to them, or the Company would not be able to carry on.

The December issue of the *Murphy Review* was the last to be published, yet to judge by its contents (with the exception of T. H. O'Brien's financial statement), nothing could be further from the editor's intentions. There was a leading article by Frank Murphy, entitled "The State or the Individual?"; a leisurely survey of liberty of conscience as seen in the early days of Rhode Island in the United States; an evocative description of a remote Irish community in "The last train from Port Clive"; the fourth in a series of articles on furniture woods, in this case on types of beech and elm, and their properties; thirteen pages of correspondence, on such topics as raising the standard of living, possible peace terms with Germany, the right of a shareholder to receive a return on his investment, the New Conception of Business as viewed by (1) Trade Union Leaders, (2) members of Methodist Guilds, (3) journalists, and two longish letters from Joan Murphy with replies from her father on whether ultimately man could have any other liberty than "liberty to serve".

Indeed, among the Notices on the back page were several which pre-supposed the continued existence of the Company—an advertisement for an assistant accountant, an announcement of the regular publishing dates for future *Murphy Reviews*, and the offer of a binding case for past issues. Other N.C.B. publications were listed as available and also names of representatives who were willing to call and explain the N.C.B. principles to anyone who wanted help to establish it in their own firms. It was announced that Kenneth Murphy had opened a Frank Murphy Hotel at Upper Colwall, Malvern, Worcestershire (Terms: £3 a week, 10s. a day, dinner, bed and breakfast, 8s.). The names and business addresses of all the local managers, sixteen in all, were published. Finally, a brief note from the editor invited readers of the *Review* to send in short essays on the meanings they attached to certain phrases in regular use namely:—

"the common good"   "an individual"
"democracy"   "national freedom"
"government"   "responsibility"

Strongly contrasting with this assumption of steady progress was a section called "The N.C.B. at Work", in which extracts from the weekly circular issued to staff by T. H. O'Brien showed just how tight the financial position was, and how the worsening national credit situation ws forcing him to adopt desperate measures.

The suppliers making the sideboards announced in the first week of November that they wanted payment within three months of delivery, not six, as originally agreed. This proposal was then withdrawn, and they stated by the end of November that they were prepared to deliver 10 sideboards in two weeks' time, then 25 a week thereafter; but the goods were to remain their property until sold by the company, when they expected their money immediately. In any case, the company had to pay within six months of delivery.

O'Brien reported that the third week of November had been "a very sticky week". He went on, "There was a minor crisis when I had to redeem a promise and stave off a Writ with post-dated cheques". On wages, he reported, "Lower scales (up to £4 10s. 0d.) have had something, with a maximum of £3. The rest of us have had to go without, and probably next week will be the same".

To get a supplier willing to take the order for 350 chairs, O'Brien, after applying in vain to finance houses all over London, had been given or promised one or two loans of £100 and £150 from staff members for this purpose which enabled him to place the order. He asked the staff to try to find sources of loans of up to £5,000 for financing more stocks of furniture.

But it was all in vain. One of the creditors decided that come what may, his firm must get what money they could, and decided to sue for it. A decision was taken to liquidate Frank Murphy Limited, and this was a tragic day for Frank Murphy and all his loyal band of colleagues. He himself had lost all his capital—he had been repaid only a fraction of what he had originally spent in the preliminary years, and that repayment had all gone in supporting his family, including two children still in full time education.

Douglas Boyd and Ronald Perry attempted to salvage what they could. They formed a new Company, "Boyd & Perry Ltd.", and sold a few sets of Murphy dining room furniture. Then the Government introduced its "Utility" furniture scheme, and all new domestic furniture had to conform to an official standard. According to Boyd, the standard was very minimal, but if the work was done to the agreed specification, it was adequate, and "far better than most of the pre-war mass market rubbish".

In May 1943 E.S.A. (Educational Supply Association Ltd.) agreed to take over the shares of Boyd & Perry Ltd., and gave the two men a three-year contract to carry out design work for them. At the end of that, Boyd and Perry left E.S.A. to form "Welwyn Woodworkers Ltd." making some school equipment and designing new furniture in laminated structures. Initially successful, the pair took on commitments too heavy for their slender capital, and eventually had to give up. Thereafter Douglas Boyd did freelance design work for various firms, and Ronald Perry settled in Canada.

Now in his late seventies, Douglas Boyd remembers with affectionate nostalgia "the wonderful days of Frank Murphy's venture into furniture. Had it not been for the shadows of the Second World War and its eventual outbreak, we might have made a go of it. I had (and have) the greatest admiration for Frank Murphy and his ability to cut through the non-essentials to the heart of the matter, and the N.C.B. was a great ideal to work for. We all suffer still from the lack of it, seen so clearly in the world of modern politics and in human relationships between the Trade Unions and employers—we haven't learned a thing".

It is fitting to close this chapter with the list of questions addressed to members of the public who opened the last issue of the *Murphy Review*;

"HAVE YOU A JOB?

If so, can you be sacked for anything except lack of integrity?

Can you use your knowledge to the full?

Have you the opportunity to learn?

Can you have full discussion with—those above you? those below you?

Are you free to do your work in the way you find to be right?

Are you treated like a human being or a machine?

Are you proud of your job?"

"ARE YOU AN EMPLOYER?

If so, do you trust your employees?

Do you discuss your business freely with them?

Are they interested in it as you are?

Do you have all the worry?

Is your business as flexible as you would like it to be?

Is it growing or standing still?

Are you proud of your business?

"If your answers to these questions are not ones you are satisfied with, it will pay you to work towards the conversion of your business to the New Conception of Business, by getting the Membership Agreement adopted."

Chapter Twelve

THE WILDERNESS YEARS

"I was ever a fighter, so—one fight more,
The best and the last!"   *Browning:* Prospice.

By February 1940 the gallant group at Frank Murphy Limited had accepted that defeat was a reality and had dispersed to make what they could of the rest of their lives. For Frank and Hilda Murphy, with all their savings gone and few opportunities open for re-employment, it was a bleak outlook. Since Frank had been in a managing director's position for ten years, the possibility in wartime of finding a Company with a suitable vacancy for his experience and ability (and with some sympathy for the ideals of the New Conception of Business) was remote.

Fortunately the liquidators could not take their home, which was rented from the Welwyn Garden City Company. Hilda therefore suggested that she should run Ludwick Corner as a guesthouse, since they had already proved that there were rooms surplus to the family requirements. She was an excellent cook, and now, through Guiding, experienced in catering for numbers; and her close friend and companion Dorothy Butterworth was willing to help. It so happened that at this time Imperial Chemical Industries Limited had evacuated staff, formerly based in the heart of London, to Welwyn Garden City, and they were finding it difficult to obtain suitable lodgings. The Welwyn Film Studios too, had visiting film stars and directors who needed temporary accommodation. So new faces were soon to be seen in the west wing of the house, while the family retreated to Mrs. Tweedie's "servants' quarters". With the money from the first paying guests, Hilda was able to order more beds and wardrobes from John Lewis Ltd., the London department store, (not, alas, from Frank Murphy Limited), and to have the three largest bedrooms partitioned so that up to ten guests could be taken at one time.

School and college fees were beyond Hilda's reach, so Maurice came home from Leighton Park School, and was taken on by Fred Palmer, a local builder and carpenter, who had been a friend of the family for several years, and keenly interested in the N.C.B. The authorities at Newnham College, reluctant to lose Joan in mid-course, found a typically British way out of the dilemma. She was awarded a major scholarship on the results of the first part of her Tripos examination, and this, with an interest-free loan from a special Student Fund, enabled her to remain at Cambridge until her Finals in June 1940.

The Battle of Britain began in the late summer of 1940, and Kenneth Murphy, who had registered as a Conscientious Objector, was allowed to offer alternative service as a voluntary ambulance driver. (The small hotel at Malvern had to be given up when the parent Company failed, and he and his wife and two young children returned to live in the Garden City). It was no easy option, and several times he was in serious danger, and on one occasion escaped a falling bomb by only seconds. He had managed to obtain work as a personnel officer in a large plastics factory, but had to travel across London to reach his work.

Meanwhile, the major creditor of the defunct furniture company was still determined to get some more of his money back, and through the liquidators issued a summons to Frank Murphy to appear in Court to say why he had not handed over to them the £1,100 he had been repaid as part of the preliminary expenses.

He refused to attend, asserting that it was unreasonable for the law to expect its citizens (whom it was supposed to protect) to attend a Court hearing in the centre of London at the height of the blitz, especially since he had all but lost a son from enemy action in that vicinity. If other Government and official bodies thought it sensible to move their activities away from what was, in effect, the front line, why should the Law Courts not do so?

Needless to say, this line of reasoning did not carry any weight with the judge, and Frank Murphy was duly found guilty of contempt of Court, and sentenced to stay in Brixton Prison until he had purged his contempt. After three weeks, a very good friend decided that enough was enough: he paid over the sum of one hundred pounds. Frank Murphy was released, and agreed to attend the Court hearing.

He wrote later:
"It was an interesting example of the fairness of the British law. The liquidators turned up in force. I was alone and had no legal Counsel. The liquidators did their best to turn it into a third degree trial, but the Court would not let them. They could ask proper questions, which I had to answer, but no more. The net result of the examination was that the Court found that I was not to blame for the failure of Frank Murphy Limited. However, one of the liquidators pursued the question of the £1,100 repayment by the directors (out of the £4,000 owing to Frank Murphy). Now, the Articles of Association of Frank Murphy Limited, actually gave me supreme executive authority, and consequently any act of the directors which I had allowed was in order; but this fact was never brought to the notice of the Judge trying the case. This was because I had no money to engage a competent Counsel. The man I was allotted looked at my case about ten minutes before going into Court, and gabbled through it, missing the essential point."

Inevitably, the liquidators won, and the Judge ordered Frank Murphy to repay the £1,100. Of course, he could only pay a tiny fraction of that amount, which he did under protest.

Later on he met the second liquidator, who privately admitted to being more and more convinced of the rightness of the New Conception of Business approach, and the injustice and inadequacy of the traditional Limited Liability Company.

Although his experience inside a British prison was a short one, Frank Murphy learned a number of things from it. In fact, he humorously suggested that everybody ought to go to prison, so that they would begin to talk about "our" prisons, not just "the" prisons, and would realise just how stupid and inefficient prisons were, if their primary purpose was to make the inmates abandon their wicked ways and lead an honest life.

Brixton Prison in wartime, had a curiously mixed population, consisting of three distinct groups:—remands, "18B's" and convicts. Most of the men on remand were not criminals but debtors who had been imprisoned for failing to keep up the separation allowances to their wives. The length of their stay depended on the amount of the arrears, and was roughly equivalent to one day for every pound owing—when the days had been served, or payment for them had been made, the man was released.

The literal interpretation of these regulations produced some comic results. For instance, if a man was not due for release until the Monday after the Cup Tie Final, he could arrange to get out to see it, providing he could pay the necessary two pounds. Even more ironic, there were cases where the wife relented after having had her husband put in jail, and came to the prison governor with money for his release. Supposing he still had seven days to serve, she then paid over seven pounds, the man was released, and the governor then solemnly handed back the seven pounds to the woman as the outstanding amount of her separation allowance.

The "18B's" were men detained under Regulation 18(b) for political reasons, having or suspected of having fascist sympathies. They were mostly followers of Sir Oswald Mosley, who was himself imprisoned under Regulation 18B. They had special privileges, smoking being a particularly valued one. Mosley, it seems, made himself very popular by secretly passing on cigarettes to the remands and the convicts, who were not allowed to smoke. But he rapidly forfeited their support by his arrogant attitude on another matter—the prison baths. Although the prison authorities provided plenty of hot water in their baths, the surroundings were considered altogether too spartan by the aristocratic Mosley and his followers. The immediate reaction of the convicts and remands was, "If the baths are good enough for us, how come they're not good enough for Mosley", and no quantity of cigarettes could restore his popularity.

The convicts were in Brixton only on a temporary basis, awaiting more permanent accommodation at places like Dartmoor. But even these, who, you might imagine, would be hardened criminals, had their streaks of honesty. On one occasion a batch of convicts was brought in, one of whom needed a fresh coat. Frank Murphy had elected to work in the prison's "Part Worn Stores" department and he handed the convict a brown coat to try on. These were used by remands and 18B's. The convict refused to take it, pointing out that he, as a convict, was allowed to wear only grey.

Another hilarious reminiscence of prison life was attendance at the Sunday church service. Frank Murphy described it thus:

"We were all shepherded into the Chapel by a Warden with a loaded rifle, and once inside, were locked in. The Warden then took up a strategic position from which he could keep an eye on all, and use his rifle effectively if anybody decided to start some monkey business. His other job was to sing the responses, which he did equally well.

I don't remember what the sermon was about but I shall never forget hearing the so-called hard-bitten convicts singing "Jesus, Lover of my Soul" at the top of their voices, and singing it very well. I realised that those fortunate enough to have good singing voices seized upon any opportunity to exercise them, not because of the sentiments expressed by the song or hymn, but just because to them singing was a real joy and pleasure. I don't blame them. If I had a good voice, I am sure I should have done the same."

Most people would have done their best to conceal the whole episode as shameful and humiliating, but not Frank Murphy. Though not glossing over the annoyance in the loss of personal freedom, and though highly critical of British penal policy as a whole, he was nevertheless struck by the common humanity of the prison inmates, sometimes pathetic, sometimes comical. To him, they were all men, no different basically from the wider community outside. He even recounted with glee how, some months later, as the Director of a newly formed Company, he paused with a companion on the steps of an exclusive hotel, recognised the commissionaire with delight as the former Warden of his "Part Worn Stores" Department at Brixton and to the acute embarrassment of his colleague proceeded to have a lovely chat about old times . . .

Back home again, he was pressed to find a job which would help Britain's war effort. Though not a pacifist, he was firmly convinced that the war would achieve nothing, and as we have seen he was only too ready to expound his views like St. Paul, in season and out of season. So when he was offered a post with the Admiralty at Haslemere in Surrey to help increase the output of radio equipment, he was unlikely to find sympathetic colleagues. Indeed, he was there only three months before those in charge politely suggested that he might be "of more use in industry", and terminated the appointment.

During his short stay at the Admiralty, Frank Murphy learnt two things. One was that each of the services was so jealous of its independence that none of them would agree to alter the specifications of pieces of their equipment, such as the jacks to a telephone operator's switchboard, to make them interchangable, and thus increase efficiency.

The second thing he learnt was that, apart from the period immediately after Dunkirk, the British war effort was barely a third of what it might have been. This was because workers on the shop floor who had initially been full of ideas and suggestions for improving output in their factories, found that they were snubbed or ignored by the management. The result was that they shrugged their shoulders and carried on doing what they were told to do, and no more. In other words, there was an enormous loss of efficiency due to the management's belief in dictating orders to the workers, instead of encouraging their initiative and enthusiasm. The managers in their turn were equally disgusted to find the Government dictating to *them*, with Sir Stafford Cripps in a position of supreme authority over industry having far more power than any Managing Director.

In Frank Murphy's view, both England and Germany believed in dictatorship, but since the Germans appeared to believe in it to a greater degree, he surmised they would be correspondingly less efficient in their war effort, and would ultimately lose the war. Such uncomfortable opinions made him a difficult colleague, so it is no wonder that his superiors at the Admiralty, whose views were very traditional, were anxious to get rid of him.

This experience, far from deterring Frank, only strengthened his resolution to propagate his views on the war effort and on the conduct of industry and government, and he spent his days back in Welwyn Garden City, typing endless letters to the Prime Minister, to the local M.P., to the local newspaper, and calling on any of his friends who still had time and patience to argue with him.

His wife Hilda had no leisure to do so; her days were fully occupied with running a busy guest house, looking after chickens and a Jersey cow, and continuing to run the local Girl Guides as District Commissioner. Frank's only contribution to the household was to stoke the boilers night and morning, and for this he was paid £2 0s. 0d. per week and full board out of the business. It is understandable that husband and wife found their interests diverging as they were less and less together, and Hilda spent more and more time with her constant companion and her loyal helper, Dorothy Butterworth.

So the domestic situation deteriorated, until in 1942 a greater tragedy brought the family together. Several cases of smallpox had been reported in the national press, and Kenneth Murphy, who had not been vaccinated as a child, thought it prudent to have this done in order to minimise the risk to his own children. Unfortunately, he was one of the rare individuals who react adversely to vaccination and in a very few days he was in a high fever, then unconscious with double pneumonia, and within a week he was dead.

This was a terrible, because totally unexpected, shock; and Kenneth's young widow, Audrey Mary, who was left with two children under the age of seven, was utterly grief-stricken, and desperate for help and sympathy. Her own mother lived two hundred miles away in Devon. Hilda Murphy, with the responsibility of the guest house on her shoulders, could spare little time to help Audrey Mary through the first most painful months of loss; and in any case Hilda was deeply reserved, and found emotional situations, especially this one when her own feelings were deeply involved, too difficult to enter. So invariably it was Frank, with unlimited leisure and the ability to talk and to listen, who spent increasing time round at Audrey's house.

The young widow was faced with the classic dilemma of one-parent families; whether to go out to work and attempt to run the home and bring up the children on the wages of a typist—then pitifully low—with the need to absent herself from work if either of the children was ill; or to stay at home and look after them on her inadequate widow's pension. Re-marriage might have been the answer, but she could not consider it in the short term. So Frank decided that he had better find a job again, so that he could contribute to their income. It had not occurred to him to get a job to support his own wife and release her from the chores of the guesthouse, but he thought her both competent and happy in her work, so no doubt he did not think it was necessary. By this time Frank had in any case decided that his marriage to Hilda was a dead letter, and all his frustrated affection and longing for unquestioning confidence in his powers had found their object; while Audrey, in her turn, found the support and comfort and loving companionship which she lacked.

The job Frank got was that of sales manager to a Hungarian

who was manufacturing and selling cigarette lighters in Britain. Her did not find it congenial to work for the Hungarian, who believed in selling to wholesalers as many lighters as they would accept, and then disregarding them and dealing direct with the retailers. He prided himself on his ability to play three games of chess at once, to which Frank's reaction was, "Yes, you clever devil, no doubt you can, but you don't know how to play cricket". He then pointed out that cricket was subject to the Rule of Law, and no one, not even the Captain, could dispute the umpire's ruling that he was out. "So to say a thing isn't cricket means that it is considered unfair, or a violation of the Rule of Law." As a dissertation on British Law was not what the Hungarian had hired him for, they very soon parted, and once more Frank turned his thoughts towards the idea of starting a business which would run on New Conception lines.

Meanwhile, life in Welwyn Garden City was becoming increasingly difficult for Hilda Murphy. She and her husband had been well known Garden City characters for years, and everyone knew that he had left her and was living in the same town with his former daughter-in-law. She felt she had to escape the gossip-mongers, and if Frank and Audrey would not move, then she would. She found a large Elizabethan manor house standing in extensive grounds just outside Sidmouth in Devon, and early in 1945 she and Dot Butterworth moved down there, hoping that their regular guests would come down to spend their holidays in the Devon countryside.

They were not disappointed. Many people booked up immediately, and some came year after year to enjoy the delights of "Knowle". It was a marvellous retreat from the nervous strain of life in wartime London, and later, from the post-war rigours of food and fuel shortages. When the lease of "Knowle" expired, Hilda bought a smaller property, "Applegarth", in the nearby village of Sidford where her Devon cream teas (by courtesy of Corduroy, the Jersey cow) became famous. But she was now in her sixties, and beginning to yearn for a quieter life. So she sold "Applegarth" and bought two old cottages in Bridge Street, Sidbury, converted them into one, and settled down to semi-retirement, having encouraged Dot Butterworth to branch out on her own at the Cherry Tree cottage by Sidbury Church (cream teas, morning coffee, and home-made cakes). That might have been the idyllic end to her life, but it was not to be. Strangely Hilda and Frank Murphy were to meet once more, in very different circumstances.

To return to Frank in Welwyn Garden City; with the war in Europe at last over, he set up another company, this time called Frank Murphy of London Limited, with the intention of making both furniture and radios. Two of his old associates from Murphy Radio and Frank Murphy Limited days were once more ready to help. They were J. Craig Robertson, who was made General Manager of the Woodware Department, and J. B. Ridley, nominated as Deputy Managing Director. Temporary offices were opened at 15 Sun Street, Hitchin, Herts, and the company began by manufacturing step ladders and ironing boards for domestic use, with the intention of branching out into radio sets as soon as the necessary licences could be obtained. The monthly diary of *FM News* (successor to the *Murphy News* and the *Murphy Review*) frankly disclosed the problems encountered by the new company.

"The timber position is the chronic irritant in this department. The Board of Trade will issue licences to us for low grade timber only, if the goods made from it are for the home market, but better quality timber is freely granted for the manufacture of goods for export. As far as the home market goes our timber buyer (Jim Shelton) has a very difficult time in finding low grade timber that is good enough for steps and ironing boards. We have always managed to keep going, apart from a few violent jerks. Nevertheless it is very disheartening to all concerned to know that we could so easily build a much bigger woodware business (a turnover of the order of £500,000 could normally be achieved without difficulty), were it not for the timber situation."

In a later reminiscence, Frank Murphy enlarged on the problems:

"We succeeded in working the sales and the output up to 400 pairs of steps per week, despite the fact that the only factory we could get had no roof (it had been blitzed), and that in order to get a roof on it we had to get a permit from the local authorities to do so. Again, we could get no woodworking machinery, either new or secondhand, without a licence, and the authorities would not give us one. However, eventually we did acquire some secondhand machinery, and got it into good working order. The only timber we could get was that recovered from air-raid shelters, full of six inch nails . . . Then, to crown everything, along came the worst winter Britain had experienced for a long time (1946-47), and with it the power shortage."

A prototype table model radio set was developed, with a proposed price of between £14 and £16 (including tax), and dealers were invited to order stocks for expected delivery early in 1947. But the old firm of Murphy Radio was not willing to allow competition from any other Company formed by Frank Murphy, and as before with Frank Murphy Limited, Murphy Radio dealers were warned not to touch the new products, or they would lose their dealership with Murphy Radio. In any case, as Frank Murphy himself acknowledged, the root cause of the failure of the post-war company was the impossibility of obtaining raw materials good enough to use, and the constant refusals of government departments to issue licences or to encourage manufacturers with initiative.

In the summer of 1947 Frank Murphy made one more attempt to get a business going in England. This time it was to be called "One World Laboratories Ltd.", and the aim was to design products such as radio sets, refrigerators, furniture etc., and also to design and exercise general supervision of the manufacturing process, publicity and distribution, through a series of operating companies in Britain and in other countries all over the world. A thoroughly impractical and grandiose scheme like this was hardly likely to attract capital and it did not. Moreover, unusually for someone who professed a belief in the Rule of Law and having everyone however important, subject to it, there was no curb on the powers of the "Chairman and Life Director" through a Board of Trustees. Indeed, there was an ominous little paragraph ascribing the failure of Frank Murphy of London Limited to "too much belief in democracy and not accepting that there was a legitimate place for force, and consequent failure to insist on complete control".

Just at this time Frank's younger son Maurice, who had been sent to Canada during the war as part of his training as an R.A.F. pilot, announced that he thought there were far better opportunities for him in Canada than in post-war Britain, and he was accordingly going to emigrate there as soon as he could. His descriptions of life over there convinced Frank Murphy that he, too, could make a fresh start in a country which seemed to welcome initiative, and was not beset with problems like post-war Britain.

So in the autumn of 1947 Frank and Audrey flew to Canada with the two children Ann and Patrick, full of hope and enthusiasm for a new life. Frank was 58, but he still had more energy and powers of concentration than many men twenty years younger.

Landing in Toronto on the last day of the Exhibition, the little party found that all the hotels within their means were fully booked, and the only room they could obtain cost them $56 a week. Eventually, however, they were able to find accommodation at $25 a month in a farm at Milton, outside Toronto, which belonged to two warm-hearted Canadians, Cedric and Laurel Harrop, and here, too, Maurice was offered a room when he was released from the Royal Air Force in November that year.

Having got a base, Frank Murphy endeavoured to start up a furniture business again, and asked Maurice to design the furniture for him. The latter's first effort was a small coffee table, which was a miniature version of the fixed dining room tables made by Sims and Wood at Ludwick Corner. Sims and Wood had been graduates from a school for cabinet makers run by Sir Gordon Russell, so the influence of that great designer and craftsman came to cross the Atlantic in this curious way.

As the farm was not suited to the manufacture of furniture, Frank Murphy rented a large old house in the main street of Milton. Apparently the house consisted of four bedrooms, a living room (which was used to store furniture in while the glue was drying), a dining room, a basement, and a great kitchen containing the typical enormous wood-burning stove which provided general warmth and hot water, and this room was for all practical purposes, the living room.

Maurice writes:
> "My task was to convert the basement into our first furniture factory. In this I was helped by Cedric Harrop, who had little to do on his farm in the wintertime. I can still remember white-washing the rough walls, stoking the coal furnace, and installing the various work-benches required.

The next step was to acquire some tools, so Father and I went to Toronto where I did the selection of the tools and Father successfully persuaded the manager of the store to give us the necessary credit. We acquired a Ford pick-up truck in which I was able to transport timber from Toronto for making the coffee tables. While I got on with making some forty or fifty of these, Father did a lot of talking in Milton and convinced a number of residents sufficiently to cause them to invest some capital in the business. My only pay was my board and lodging.

Next, I began making prototype chairs similar to those that had been made in England. Father found a firm in Toronto to produce the plywood seats shaped to fit the occupant's bottom, although they did not achieve the original depth of depression of Douglas Boyd's chairs. They could not, however, have been too uncomfortable, as a representative of Simpson's, a large store in Toronto, became interested in making these chairs under contract. This scheme never materialised, probably because of Father's lack of capital, or perhaps because he preferred to go it alone and be independent of Simpson's."

It was one thing to make some sturdy and attractive coffee tables; but it was quite another to try to sell them to Canadians, whose lifestyle at that time was entirely different from middle-class Europeans. Frank was still vainly trying to get orders for these, when fate struck them a severe blow. Audrey Murphy fell ill, and had to go into hospital in Hamilton for a major surgical operation. There was no money to pay the surgeon's fees, so Maurice was despatched to Hamilton with a truckload of coffee tables equal to the value of his bill. As Maurice said, "I can remember carrying these, at least twenty of them, in their cardboard cartons, and dumping them outside the surgeon's office right in the hospital. I didn't see him, but the poor fellow must have got an awful shock when he returned from his rounds."

Within a month it was obvious that the business was going to fail. For once, Frank's courage in the face of adversity deserted him. He suffered a kind of nervous breakdown, and tried to release his terrible frustration by throwing the crockery at the kitchen walls, and finally stumped out without saying where he was going. Audrey was terrified that he was going to throw himself on the nearby railway track and commit suicide, so Maurice rushed out in the truck to look for him, searching all the town and the countryside. When darkness fell, he returned home fearfully to the house, only to find his father sitting in the kitchen in a much calmer frame of mind.

Soon after this, the tools and the truck were re-possessed because of failure to make the required payments. Frank and Audrey decided they must return to Toronto and find separate jobs to enable them to make a living and to support the children. Maurice was rescued by Cedric and Laurel Harrop, who took him on as their "hired hand" at the farm. Though extremely grateful to them, he felt he was not cut out to be a farmer by profession and when he heard that the Royal Canadian Air Force were starting to recruit pilots again as part of their post-war programme, he spent his last $10 on a trip to Toronto to be interviewed.

Meanwhile Audrey managed to find herself an office job in Toronto, but Frank did not find it so easy. His creed was "I can be either at the top or at the bottom, and nowhere else". Putting this literally into practice, he first tried driving a taxi; but his age, the unfamiliar streets, the traffic, and perhaps also the emotional and financial strains of the past months made him accident-prone, despite the fact that in the past he had been considered an excellent driver. He then applied for a post as hotel porter and odd job man, for which he was paid $20 and his keep, on one of the summer hotels on Toronto Island. According to Maurice, who visited him there, he seemed quite happy and gaily recounted his experiences, mostly about the guests, whom he intrigued by means of his lengthy discourses on how to run a business.

His marriage to Audrey, which had taken place in Canada on December 2nd 1947, had by now been annulled. Though Audrey still regarded him with affection and respect, she recognised that the 26-year age-gap between them was too great to be bridged, and she felt an intense longing to go home to England. She wrote to her father, who forwarded the passage-money for the return trip for herself and the children, and she booked their passages on the liner "Franconia", due to sail in October 1949.

News of the unusual porter, or janitor, at Hotel Manitou on Centre Island must have leaked back to the city, for early in October 1949 a Canadian reporter, James Cooper, sent an article to the *Daily Express* in England, headed *"Murphy (remember him) is a porter"*. The article briefly described Frank Murphy's rise to fame with Murphy Radio as "The Man with a Pipe.", the formation and failure of the furniture venture in England, his emigration to Canada and attempt to set himself up as a business consultant, the recent experience as a taxi-driver, and now as a hotel porter. It ended:

"His wife and family are on the way back to Britain, but Frank Murphy means to stay on Centre Island until the water freezes and the season ends. He has a philosophy. 'So long as I have my pipe and a few ounces of tobacco, I'm happy', he says . . ."

This stung Audrey into writing a strong letter in reply, which the *Daily Express* also published, to its credit, a week later. It deserves quoting in full:

"I do hope that great Canadian, Lord Beaverbrook, will allow his paper to print the other side to the story 'Murphy is now a porter'.

Actually since the date of the interview Frank Murphy has been promoted, in a way, to maintenance man. You see, with a people as broad-minded and generous as most Canadians, they are very quick to realise that a man who has the letters M.I.E.E., B.Sc., M.B.E. after his name—a name, which as you say, founded a radio firm which is still proud to use it and which from 1930 onwards gave employment to hundreds of citizens, as well as a great boost to Welwyn Garden City—must have a great deal of know-how to pass on.

Your report did not mention that most evenings Frank can be seen in the hotel lounge discussing most subjects from art to international affairs and recently, your own sorry dollar plight, with many prominent citizens of Toronto. Although they may call him 'Frank', equally he calls them Bill, Ted or Harry.

In England, I know, it is difficult to conceive such a fantastic state of affairs, but here in Canada a man is judged rather by his character and knowledge than by how much he has in the bank or of what firm he is president.

No, Mr. Editor, I was on the point of bringing Ann, 14, and Patrick, 11, back to England. I had the tickets to sail on October 25th, but just in time I read your article and came to my senses.

I knew then that never again could I bring my children in that class-conscious—and as the *New York Times* calls it—'too tight little island' called England. Do not think I do not love England; it is still my home and I will cherish memories of its green and beautiful countryside until I die. But in the North American Continent I have found a land like the pilgrims before me, where the sons and daughters of porters, garbage men, street-car conductors, labourers, and any man who toils honestly with his hands rather than with his brains, may go to universities and colleges and emerge lawyers, judges, politicians, and leaders of their own virile and glorious lands. I know we have had our rough times in the past two years. One does not come to a new country and expect to be handed security on a plate. But we have also had a terrific amount of fun.

And so back to that once eminent industrialist of England, Frank Murphy, now porter and maintenance man, and to all men who raise the standard of human dignity, I raise my glass. Please God there will always be men and women like him."

AUDREY MARY MURPHY, Lakeshore Avenue,
Hanlans Point, Toronto.

Even allowing for some editorial tampering so as to popularise the letter with *Express* readers, this was still a good letter, and its message was clear.

At this time Audrey and the children were staying in two small rooms at the Manitou Hotel, and she was crossing the lake daily to commute to work by ferry. The owner of the hotel, Bill Sutherland, introduced her one morning to a fellow commuter, Norman Hand, and they soon became friends. On November 10th Norman asked Audrey to marry him, and she accepted, and cancelled the passage home to England. They were married on January 28th, 1950, and lived a close and happy life until Norman's death in 1979. Recently she wrote, "Frank had a brilliant mind, great human characteristics, a lovely sense of humour and a loving heart. From the time I first knew him, (April 1932) he was a lonely man who with the coming years would become ever more lonely. He was in love with me in 1942 and I loved him. But it all fell apart in a young country with young people, and the

26-year gap became overwhelming". She saw Frank only once more after her marriage to Norman, but it was she who, when Frank died, made arrangements for his funeral.

The job as hotel porter having come to an end, Frank looked around in Toronto and took a job as a guillotine operator in the printing department of a book club. Although the pay was not high, it was increased to $35 a week after three months, when the manager promoted him to foreman. Frank was obviously handling his duties competently, but in his spare time he committed his philosophy of life to paper in a booklet entitled "The Root of the Matter", and when his somewhat unorthodox views were noted by one of the directors of the firm, Frank's boss was under pressure to dismiss him. Rather than cause a fair and kindly man embarrassment, Frank resigned, and once more was without a job.

He thought it over and finally came to the conclusion that he would continue to fail unless he tackled his "real job". That was "to establish a business which was a real democracy". Once more he embarked on a radio business, and found some backers to put up the capital for the *Murphy Partnership*. Maurice remembers visiting a tiny factory in North Toronto, and meeting one of the backers. Maurice himself had persuaded a young English teacher, Carol Maddock, to visit him in Canada, and in December 1949 they were married in the RCAF chapel at Rockcliffe. He and Carol later bought one of the radiograms designed by Frank (which they still possess) as well as the original wiring diagram for the set.

The Murphy radiogram, or radio-phonograph, as it was called in Canada, was being produced by Triton Electronics Limited, a company owned and controlled by the Murphy Partners, at Bloor, Toronto, and in the accompanying literature there occur several very familiar Murphy features. For instance:

"In designing the 'Murphy' radio-phonograph, I tried to design something which was *useful, reliable,* and *good value for money* . . ." . . .

"Now every object has many uses, but of all of these uses, one is the primary one. For example, the primary use of a chair is to sit on it and to do so in comfort. Similarly, the primary use of a radio-phonograph is to be entertained through listening to it. It may also serve as a support for a vase of flowers, but this certainly is not its primary use, nor one calculated to improve its performance in respect of its primary use." . . .

"If we sold our radio-phonographs so that they could be inspected at every store selling radio, we should have to increase the price to something like $475; in other words, the extra convenience would cost each customer about $200."

Here is another paragraph with the authentic Murphy ring:—

"Why, then, is the quality of reproduction of a 'Murphy' above average, seeing that it uses well-known and ordinary circuits? The answer is in the same category as the answer to the question, 'Why do some restaurants make a decent cup of coffee, and lots of others don't?' The art of making coffee is well-known and anyone who follows the established sound lines for making coffee can and will make a good cup of coffee, provided they can afford to buy good coffee beans to start with."

and finally:

"We try to make a 'Murphy' as reliable as possible by using well-made components and by careful workmanship, but being pessimists in this matter of reliability, we make the chassis (radio tuner and audio amplifier) extremely easy to remove from the cabinet. We also lay out the chassis in such a way that all parts are readily accessible. With many sets it is usual to warn the service man to replace all wires precisely where he found them. No such warning is necessary with a 'Murphy'. From a wiring point of view, the set is virtually fool-proof."

"*Guarantee*: If it is our fault, we put the trouble right. In case of doubt, we give the customer the benefit."

Ten pages of detailed explanation of the set's design, and the reasons why each part of it was so designed, were followed by two brief pages summarising the aims and objects of the Murphy Partnership.

By this time Frank Murphy had managed to compress his ideas on industry into a single theme—that of "fairness". He had found that this concept was simple enough to be understood by all, whatever their background or standard of education. So the first page on the *Murphy Partnership* was headed with the question "What is it?" with the answer:

"An association of men and women who accept the rule of 'fairness' in their own lives; that is, those who act on the basis of 'That's fair, that settles it'." The next question is obviously, "What do they mean by 'fairness'?" The reply is no surprise to those who have followed Frank Murphy's career:
"Being governed in their actions by the three principles:—
No rights without obligations
No man to be judge in his own cause
All men to be equal before the law."

The reason given for this is "because they are satisfied that the histories of the British Commonwealth and the U.S.A. prove that 'playing the game' is the way to secure individual liberty and the best standards of living for all". The rule of "fairness" for capital, labour and consumer alike, is contrasted with contemporary business, whether that of a Limited Liability Company, a Trade Union or a Consumers' Co-operative, all of which, he asserts, are governed as a dictatorship, on the principle that the might of capital, of labour, or of the consumer as the case may be, gives them the right to dictate to all others.

He then lists the advantages of governing business by the rule of "fairness":

"It gives to everyone engaged in the business, whether providing capital or labour, the maximum freedom to express their own individuality, but only in the service of others.

It gives to everyone, whether capital, labour or consumer, the maximum assurance of a square deal, because all engaged in the business have the maximum freedom to be honest, to be fair—in fact, are given every encouragement to be so.

It gives higher standards of living; to capital by way of dividends, to labour via wages, and to consumers through better value for money, because of its higher efficiency and therefore greater output per man/hour.

It eliminates strikes and lock-outs, because both capital and labour, being governed by 'no man to be judge in his own cause', accept that disputes which they cannot settle themselves shall be settled by arbitration.

It eliminates the risk that our civil lives shall be governed by a dictatorship and our present democratic way of life suppressed, because as Abraham Lincoln said, 'two opposed forms of government cannot continue to exist side by side'. Either the present dictatorship government will destroy and replace the democratic government of our civil lives, or we shall find a way, as we Murphy Partners believe we have, to govern business democratically, and then make the necessary effort and take the necessary risks to change the government of business from dictatorship to the rule of 'fairness'. . . ."

As in the case of the English furniture company, Frank Murphy proposed that all who joined the Partnership should sign an Agreement accepting the principle of "fairness", and further, that democratic control should be established by making everyone who signed a Partner, whether providing capital or labour, or both. Every Partner was to be entitled to one vote, which could be exercised by secret ballot, and all must be kept informed of the activities of their Partnership (no doubt by means of a monthly news sheet).

The final paragraph shows that Frank Murphy did not equate "fairness for all" with government by endless committees:—

The Partnership proposes to achieve its objective . . ."by entrusting the management of its affairs to a General Manager, elected by a majority of the Partners, and chosen for his proven practice of fairness and his proven business ability".

For a couple of years between 1951 and 1953 the new company or partnership seemed to be making reasonable progress, more so after July 1952, when Frank assumed complete managerial control; and once more he was happy to be working with congenial colleagues and producing a sound engineering product. He kept in touch with Maurice and Carol, and with his grandchildren Ann and Patrick, on whose education he still kept a benevolent eye. Otherwise, he led a rather lonely life in rooms in Wellesly Street in Toronto.

In 1953 Maurice invited his mother, Hilda Murphy, to visit him and Carol at Trenton, the RCAF base where they were then living. They took her to Toronto, had lunch with Ann on the Island, and later met Frank. All four went to see the film of "An American in Paris" and had a very enjoyable time, and for a short time Hilda and Frank were left alone together.

One can imagine Hilda's very mixed feelings at this interview, but Frank was overjoyed to see his old partner again. After four years on his own, he was beginning to realise his deep desire for companionship and mutual affection, and he ended by asking her to give up her life in England, to marry him again and settle down in Canada.

Realising that it was a desperately hard decision for her to take, he suggested that she return to England and think it over, and if, after six months he could be certain of offering her a secure home and future, he hoped they would once more come together.

Hilda's pride had been deeply wounded when Frank left her, but that and her natural reserve had enabled her to make a new life for herself without him. Now, the whole painful episode was recalled, and threatened to upset her hard-won equilibrium. Frank had declared that he loved her and wanted her back, and she had to admit that, in her heart of hearts, she had never loved any other person as much. He obviously needed her, and she knew that she could give him the same kind of vital support and affection as she had in their early days of marriage. But they were now both older, over sixty, in fact, and she doubted whether, after so many years of depressing reverses, Frank could recapture his former astounding energy, and make a success of his new business. Besides, she had other responsibilities. First, there was Dorothy Butterworth, who had been her loyal and affectionate colleague through the tough years of establishing the guest house business, and who now, though encouraged to achieve independence in her own little business, would still miss Hilda tremendously if she went to Canada. There was also Joan, whose own marriage had foundered during the war, and who was struggling to bring up two boys and give them a good education. In the midst of her mental and emotional upheavals, Hilda must have remembered with a wry smile Frank's oft repeated dictum of the old days, "When in doubt, don't!"

The decision was not made any easier for her early in 1954, when she learned that Frank's radio business had failed. He, however, was undismayed, and in a personal letter to Hilda, written on 2nd July 1954, he thanked her for the "two very nice pairs of socks which came yesterday" (doubtless a birthday present): and discusses the pros and cons of getting a Canadian called Rupke to put up $5000 to start another business, or alternatively, taking a sales manager's job with the firm of Philips.

"As I have said before and with which I am sure you agree, that if each one of us can succeed in being fair ourselves, then we can call the fairness side a day. The other side is that one wants to express oneself socially in one's work, that is, to do something which would prove of real benefit to others, and in my original thinking about firms like Philips, there is so much hanky-panky about their sales method that I did not think this possible. But I have changed my mind. When I left Murphy Radio I set out to find one thing, but actually, I have found out two. I set out to find what were the right human relations in industry and how to achieve them. I consider I have solved this problem. But I have also found out what is wrong with the existing methods of selling mass-produced goods, and what to do to put them right. I expect you will remember that I always considered that I did the British public a lot of good because my method of selling Murphy sets, first, directly saved the public a lot of money, and then indirectly, even larger sums through forcing my competitors to give better value. I can do the same now to the public all over the world through a firm like Philips. I estimate that if Philips were to adopt my selling ideas, not only would their sales go up by leaps and bounds, but in due course, in their TV buying alone, the Canadian public might save as much as $100,000,000 per annum.

So whether it is the Rupke road or the Philips one, I am satisfied that I can satisfy both my personal desire to be fair and also the social side of my nature which requires that I should do something helpful to my fellow man. I think that one of my snags has been that I am an engineer by training and a reasonably good one at that, and all engineers dislike the idea of saying goodbye to engineering and going over to the sales side. I know in starting M.R. I took on the sales side because it was so obvious that Ted was not competent to do so, and I always felt that life was a bit tough that I should be forced to give up engineering. I don't think so now, because I realise that I have more to offer the world with my sales knowledge than I have with my electronic. I also have a hunch that many of the scruffy engineering practices will disappear when the selling side is put on an honest and workmanlike basis.

Well, time marches on, and the tide and postman wait for no man, so if I am going to catch the 5 o'clock post, I will have to stop. There is only one more thing to say, that is—I love you.

FRANK.

P.S. If I say it often enough the odds are you will realise that I do love you."

It was in a way fortunate for Frank that neither Philips nor Rupke took to his ideas, because he at last found an opening where he could be of service to others and also pass on some of that valuable engineering knowledge. He was taken on as a mathematics teacher at the Collegiate School in Bloor, a suburb of Toronto. Here he was happy, and it seems well liked and respected by both his students and his fellow teachers. I wonder how often he was side-tracked by ingenious students into reminiscences about his former experiences; but there is no doubt that by temperament he was always a brilliant teacher.

On the morning of 23rd January 1955, Frank Murphy arrived to take his first class of the day and had just begun work, when he collapsed with a severe heart attack. He was taken immediately to Toronto Western Hospital, and according to the doctor attending him, became conscious for a few moments. "What am I doing here?" were his last words.*

The doctor afterwards told Maurice that he thought his father was unaware of what was happening, and that he had suffered no serious pain. It is possible that, like various other members of the Murphy family, he suffered from angina, and had mistaken it for severe indigestion. Throughout his life he had always enjoyed excellent health, and no doubt it never entered his head that his life was in danger. He was only 65.

Maurice was with the Canadian Air Force in France at the time of his father's death, but was flown back to Canada within two days, in time to attend the funeral. In addition to Audrey, Ann, Patrick and himself, he was surprised to see a large contingent of teacher and students from the Bloor School there, and afterwards they spoke warmly to him of their respect and affection for his father.

*Frank's grand-daughter Ann was summoned by the doctor and arrived within half an hour, only to find that Frank had already died.

On returning to Frank's apartment, Maurice found very little of real value, except the famous gold watch which the Northern Murphy Dealers had given him in 1937. This he sent to Hilda. The bank balance of $343 sufficed for the funeral and cremation expenses; and the few remaining radio materials, which Frank had intended to use for research, were sold off.

When news filtered through from Canada of Frank Murphy's death, Jack Hum, who was by now editor of the *Murphy News* in place of the late Stan Willby, wrote a brief, carefully-worded obituary:—

FRANK MURPHY 1889-1955

We have to announce with regret the sudden death last month in Toronto of Mr. Frank Murphy, who was a cofounder with Mr. E. J. Power of Murphy Radio Limited. The news of his passing first became known when it was cabled to *The Wireless World* by Mr. P. G. A. H. Voigt, the well-known radio authority, now resident in Canada, who added that he was attending the funeral on 29 January.

Mr. Murphy, who was born in 1889, was with Murphy Radio for the comparatively short period of seven years; but they were years of fundamental change in the radio industry, both in respect of design and manufacture and of distribution. The problems of effective and economical distribution in fact interested him possibly more than anything else. His conception of a limited manufacturer-to-dealer policy, while not new, came at a time when something like it was badly needed in British radio distribution.

By the middle 1930's when Murphy Radio had grown to be a major unit in the industry, it seemed that Frank Murphy's restless mind was in search of quite different worlds to explore. His interest in furniture manufacture took him to the United States in 1936, and he came back with many new ideas. In recent years he had lived in Canada.

Frank Murphy was a complex person, and while it is nearly twenty years since he exerted any impact on the Murphy Radio organisation, his passing will sadden the many who still remember him with affection.

CONCLUSION

So ended the life of that extraordinary man, Frank Murphy. A brilliant engineer, a highly successful entrepreneur, a charismatic leader, a kindly teacher, an original thinker on human relations in industry, an affectionate parent, an exasperating correspondent—he was all of these things. To many who knew him, his last years reflected only failure, yet that was not how he himself regarded them. In 1953 he wrote, "I feel that if we learn at all, it is through our failures. In general, our successes teach us nothing, because while everything is going smoothly, nothing is forced to our notice; but with a failure, the difference between what we hoped for and what we have achieved is so blatant that we cannot help noticing it . . . I think it can be said that the cause of all our failures is ignorance. Consequently as we become more knowledgeable and less ignorant, we shall have fewer failures, but we shall never achieve the perfect state of no failures, at any rate not this side of the millenium. As the Latin tag says, 'To err is human . . .'."

The fact that the most successful company he founded, Murphy Radio, has now finished its existence, would not have troubled Frank Murphy. He would probably have pointed out that it disappeared because other companies were able to give greater value for money to the public. He left Murphy Radio because an ordinary limited liability company could not allow its workers and shareholders to express their individuality in the service of society. He left the country of his birth, because he thought that post-war Britain did not believe in the freedom of its citizens to express their individuality. He felt, and I think rightly, that if people spend most of their lives in a work situation which is based on dictatorship—however benevolent—then eventually the attitudes they adopt at work will seep through to wider spheres of national and international government, and ultimately the centuries-old right of individual liberty which has characterised the western democracies will disappear.

On the other hand, he would have preferred to put it more positively. If people come to their senses and recognise that freedom is indivisible; if they see that changing the outlook of employers, capitalists, trade unionists and consumers to accepting that "no one has rights without obligations", "no man should be judge in his own cause", and "all must be equal before the law" are the fundamental principles of industrial as well as civil life: that would release such an enormous tide of energy and enthusiasm that our lives would be transformed. Strikes, lockouts, trade union and government restrictive practices, all would be seen as the wrong answers to the wrong questions. When we are no longer concerned with who is to have the most power, we shall not act, as at present, on the theory "Might is right". The greatest will be the one who is giving the most service to society, and giving freedom to others to do the same.

APPENDIX 1 - THE MURPHY FAMILY TREE
(Persons mentioned in this biography)

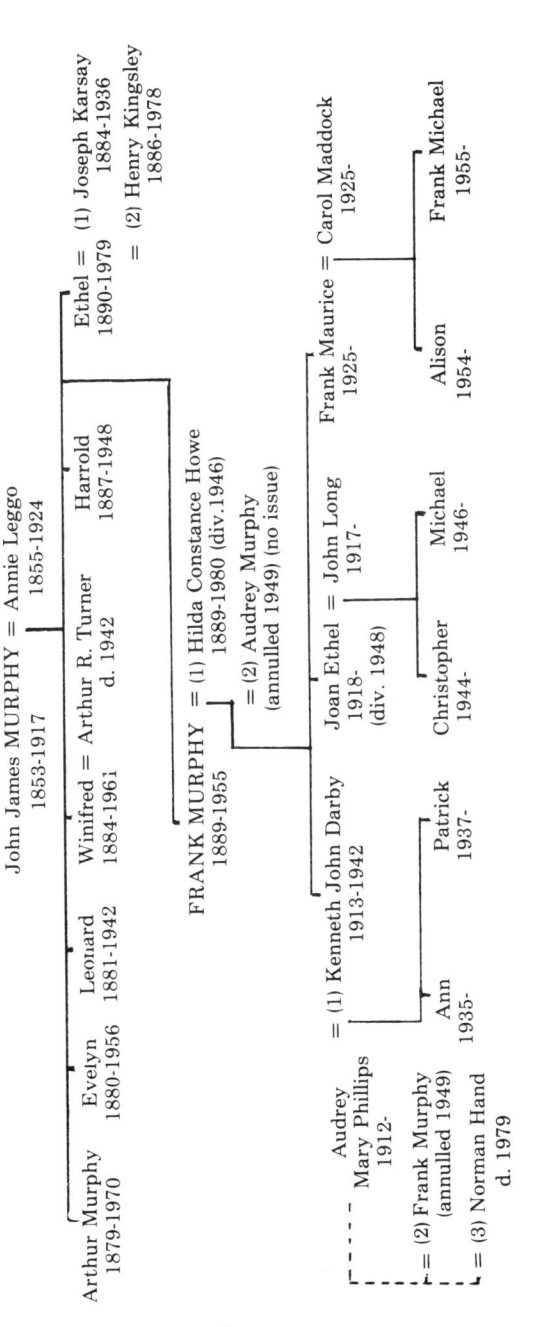

John James MURPHY = Annie Leggo
1853-1917   1855-1924

Arthur Murphy Evelyn   Leonard   Winifred = Arthur R. Turner Harrold   Ethel   = (1) Joseph Karsay
1879-1970   1880-1956   1881-1942   1884-1961   d. 1942   1887-1948   1890-1979   1884-1936
                                                                                          = (2) Henry Kingsley
                                                                                                1886-1978

FRANK MURPHY = (1) Hilda Constance Howe
1889-1955   1889-1980 (div.1946)
             = (2) Audrey Murphy
                   (annulled 1949) (no issue)

            = (1) Kenneth John Darby   Joan Ethel = John Long   Frank Maurice = Carol Maddock
                  1913-1942   1918-   1917-   1925-   1925-
                                          (div. 1948)

Audrey   Ann   Patrick   Christopher Michael   Alison   Frank Michael
Mary Phillips   1935-   1937-   1944-   1946-   1954-   1955-
1912-
= (2) Frank Murphy
      (annulled 1949)
= (3) Norman Hand
      d. 1979

Appendix 2

ARTICLES AND BOOKLETS WRITTEN BY FRANK MURPHY

Making Wireless Simple		1930
Some Notes on the Design of a Battery-operated Portable		1930

Murphy News articles:

Freedom of Choice	Aug.	1934
The technique of buying and selling	Sept.	1934
How should distribution be limited?	Nov.	1934
Replies to critics of price-changes, etc.	Mar.	1935
The Rent Theory	June	1935
The best way to run your business	Nov.	1935
What is a Shop?	Jan.	1936
How I came to start Murphy Radio	Nov.	1935
27 more thoughts on Retailing	Apr.	1936
Free Speech	June	1936
Government in Industry	July	1936

Murphy Retail Turnover for 1934 and the Population served per Murphy dealer's shop	April	1935
Some Arguments on Advertising (and Murphy Advertising in particular)	July	1935
Some Thoughts on Retailing and Distribution	April	1935

Six preliminary booklets on "THE NEW CONCEPTION OF BUSINESS" (Nos. 1, 3 and 4 missing)

No. 2 Ideal, Commonly Valued End, and suggested organisation	June	1937
No. 5 This Freedom	Dec.	1937
No. 6 The Constitution	Feb.	1938

Murphy Review articles:

The Aim, Policy and Organisation of the Company	Sept.	1938
Aesthetics (the artist and the engineer)	Oct.	1938

APPENDIX 2 (continued)

Is resisting Hitler more important than establishing the N.C.B.?	Oct.	1938
Good Beds	Oct.	1938
Unemployment	Nov.	1938
How we are getting on	Nov.	1938
The practical and the aesthetic	Nov.	1938
Hire Purchase	Nov.	1938
God in Business	Nov.	1938
Should Capital be paid?	Dec.	1938
The Churches and the N.C.B.	Dec.	1938
Shareholding Members to sign the Agreement	Feb.	1939
The Working Man and Liberty of Conscience	Apr.	1939
Progress Report on Frank Murphy Ltd. (given at a public meeting)	Apr.	1939
Chairs and Factories (Frank Murphy Ltd. planning to establish its own factories)	July	1939
The Next Step (mergers with interested dealers)	July	1939
Civil Disobedience and Liberty of Conscience	Sept.	1939
Peace Terms	Oct.	1939
Business as usual; the immediate aims of F.M. Ltd.	Oct.	1939
The Real Job—to Live, and Live Better	Nov.	1939
The State and the Individual	Dec.	1939
Liberty of Conscience and Liberty to Serve	Dec.	1939
THE ROOT OF THE MATTER—thoughts on ethics, religion, freedom, evolution, laws and industry (written in Canada)		1950
A SHORT AUTOBIOGRAPHY		1953
The Murphy Phonogram		1953
Principles of the Murphy Partnership		1953

Appendix 3

EMPLOYMENT AT MURPHY RADIO 1932-36

Average number of people employed in the works (in Sept. 1929, only 7)

Year	Number
1932	430
1933	495
1934	515
1935	640
1936	759

Annual turnover per person employed

1932	*£1260*
1933	*£1490*
1934	*£1630*
1935	*£1650*
1936	*£1320*

Ratio of operatives to non-production staff (e.g. office, research, maintenance & management).

Year		%
1932	249	58
1933	281	57
1934	253	49
1935	320	50
1936	371	49

Standard working week—46¾ hours (no Saturday working)

Standard wage rate for men over 21   *for women over 21*

	per hour	per week	per hour	per week
1932	10d.	38/6d.	6d.	2351½d.
1933	1/-	46/3d.	Not Available	
1934	1/1d.	50/1d.	Not Available	
1935	1/2¾d.	54/11d.	8¾d.	31/10d.
1936	1/2¾d.	54/11d.	8¾d.	31/10d.

Female labour employed as percentage of total operatives

Year	Average total	% of total operatives
1932	74	19.4
1933	60	13.5
1934	38	9.7
1935	53	11.1
1936	55	10.0

(As a rule, the Company preferred to employ male operatives, considering it "socially important that actual or potential heads of families should be given preference, where there is regular and suitable work available for them.")

Stability of employment

Year	Total leaving (a)	Total employed (b)	(a) as % of (b)
1932	511	1024	49.9
1933	443	820	54.1
1934	329	739	44.5
1935	395	939	42.0
1936	248	783	31.7

Analysis of males leaving

Year	Discharged for unsuitability	Left of own accord	Discharged due to seasonal variation in sales
1932	17.6	13.4	71.8
1933	7.65	11.7	59.5
1934	12.0	18.0	43.8
1935	5.65	18.8	50.0
1936	0.55	12.4	27.6

Working Hours

To obviate sacking a large number of workers immediately after Christmas, the Company tried to keep a fairly regular workforce employed through the year, but on a varying basis, as follows:

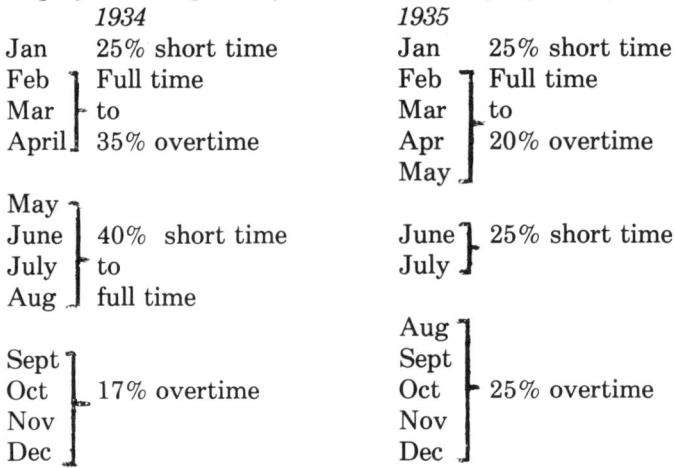

1934		1935	
Jan	25% short time	Jan	25% short time
Feb ⎤	Full time	Feb ⎤	Full time
Mar ⎬	to	Mar ⎥	to
April ⎦	35% overtime	Apr ⎥	20% overtime
		May ⎦	
May ⎤		June ⎤	25% short time
June ⎥	40% short time	July ⎦	
July ⎬	to		
Aug ⎦	full time		
		Aug ⎤	
Sept ⎤		Sept ⎥	
Oct ⎥	17% overtime	Oct ⎬	25% overtime
Nov ⎥		Nov ⎥	
Dec ⎦		Dec ⎦	

Expected working hours during 1936
Jan, Feb and Mar - Full time
Apr, May and June - 10% short time
July and August - 10% short time
Sept to December - 15% overtime

The Company's ideal was to provide stable employment for 52 weeks during the year. Methods of achieving this:

(a) by persuading the public to buy Murphy sets at an even rate throughout the year (N.B. "Buy your Murphy Now" campaign in 1933 & 1934).

(b) by "even load" manufacturing to build up stocks during the summer for anticipated rush of pre-Christmas sales.

(c) by getting dealers to order and stock up in advance (the Even Stocking Plan), so as to prevent the works being clogged with unsold sets during the summer months.

Holidays with pay.
From 1936, operative were paid for two weeks' holiday after six months' employment (or for one week's holiday after less than six months' employment), as well as for the statutory six bank holidays. Prior to this, over a year's continuous employment with the Company was necessary qualification for a holiday with pay.
Sick leave.
Full pay during sickness, less National Health Insurance benefits, paid from April 1936.
Canteen. Hot 2-course lunches available at 9d. per head in 1936. The publication of the above statistics in the *Murphy News* issue of April 3rd 1937, prompted a number of appreciative comments from the staff, the only criticism being on the subject of the need for equal pay for women workers doing the same jobs as men. Mr. A. Jeffrey, of No. 2 shop, wrote as follows:

"The advances in employment conditions have been consistent, soundly based, and scientifically effected as clearly demonstrated in the considerable data shown, but I cannot resist feeling that the following facts might have been included in order to illustrate to the full the value of the progress made.

(1) During the period covered by the review, the nation passed through a wave of unprecedented trade depression, when values and standards fell to chaos, the labour market was glutted, and the watchword "cut everything and everybody to the last thread".

(2) The progress quoted did not arise through any dissatisfaction of the employees of Murphy Radio over their conditions of employment but was rather the outcome of the generous and public-spirited policy of the directors to make employment within the firm really attractive.

(3) The period dealt with in the review was virtually the total lifetime of the company in the radio market, and it is seldom indeed that one hears of such consideration for employees on the part of a firm engaged in the initial battles for a place in the trade.

(4) The review is almost stupidly modest about the progress achieved, makes no claim to having set up new standards in employment, merely remarking that particulars of other firms' conditions were not available. A

bold challenge might have been made here, in comparing employment by Murphy Radio with employment by many other radio firms, whose factories throughout their history have been little more than casual wards.

To Murphy Radio has now passed the mantle of *Premier Radio Employers* and with it the initiative and responsibility of setting the highest standards the trade can justly allow."

INDEX

Admiralty, F.M.'s wartime post in, 172
Advertising, the function of, (C. R. Casson), 42
Advertising, distinctive style of Murphy, 34-51
Alcock, John (Trustee of Frank Murphy Ltd.), 133
Articles of Association of Frank Murphy Ltd., 170
Automatic Tuning Correction, 56
Automatic Volume Control, (in Murphy A8), 53
Avery, Bernard (N.W. London Murphy dealer), 105

Baker, Geoffrey, 105
Barnes, William, letter to F.M., 43-45
Battery-operated Portable, Some Notes on the Design of a, 30-31
Beds, Good, 145-146
Berry, George, 83
Birtles, (Sheffield Murphy dealer), 107
Blenheim Barracks, 17
Bloor Collegiate School, Toronto, 189
Boyd, I. D. A. (Douglas), 138, 141-145, 166
Branches, Murphy, or Murphy Dealers, 88-89, 108
Briggs, Susan ("These Radio Times"), 23, 45
Brixton Prison, 170-172
Broadcaster, The, F.M.'s article in, 86, 101
Bromley, Ted (Cowling Bros. of Leicester), Acknowledgments
Bush Radio, 87, 123
Butler, ("W. H. B."), 151
Butterworth, Dorothy, 168, 174, 175, 187
Butterworth, R. J. (journey through Spain during Civil War), 113

Cabinet Design of Murphy Radio sets, 58-72
Canada, emigration to, 178
Carne, Sidney (first Works Manager of Murphy Radio), 32, 95
Cassandra, of *Daily Mirror,* reply to Bed Questionnaire, 147-148
Casson, Charles Rupert, 20, 34-45, 48, 49, 61, 77, 111
Casson, C. R., Ltd. (formerly Murphy-Casson Ltd.), 41
Cavalcade magazine, article on F.M., 52
Chair, Dining, designed by J. D. A. Boyd, 153-155
Coffee tables, made in Canada, 178
Colebourn, T. H. (I.O.M. Radio Dealer), 78
Coulton, Arthur (Rossendale Murphy Radio Dealer), 96
Consumer dividend, proposed, for Frank Murphy Ltd., 158
Consumer Research, reliance on, 147, 152, 156. See also P. K. O'Brien
Cussins & Light (York Murphy Dealers), 82-83
Cussins, Denys, 83-84

201

Daily Express, Letter to, written by Audrey Murphy, 181
Dealers, Murphy, sense of brotherhood, 81
Dealers' Associations, Murphy, 91
Dealers' Meetings, Murphy, 77, 79
Denton, George, 77, 87
Design of Murphy Radio cabinets, 39-49; of Frank Murphy table, 151-153; of Frank Murphy chair, 153-155; of Frank Murphy sideboards, 153
Dining chair, how evolved, 153-155
Discount Dealers'—reduced to 27½%, 76; restored to 30%, 114
Dealers, Murphy, 73-92
Distribution, Theories of, 85-87
Dividend, Proposed consumer (Frank Murphy Ltd.), 158
Dobson, Geoffrey (Huddersfield Dealer), 90-91
Drazin, I. (Hampstead Dealer), 99
Drew, Roy (Croydon Dealer), 58
Drysdale, J. D. (Murphy representative), 95

Ediswan Company, 21
Educational Supply Association Ltd., 156
Employment, Conditions of, at Murphy Radio between 1932 to 1936, Appendix 2
Engineer's approach to design, 61, 135
Engineering Publicity Service Ltd., 20
Even Load Plan, 118
Express, Daily, Letter to, 181

Farnborough, Hants., Royal Flying Corps Wireless School at, 17
Fenn, Rev., R. E. (Trustee of Frank Murphy Ltd.), 133
Filene, Edward A. ("Successful Living in this Machine Age"), 125
FM News, 176
Folkestone-Cologne Aerial Post, 17
Ford, Henry, influence on Frank Murphy, 23
France, service in R.F.C. 1917-18, 16
Frank Murphy Ltd. (first company based on F.M.'s "New Conception of Business"), 134-167
Frank Murphy of London Ltd. (1946), 176
Freedom of Speech, 102-103, 105, 110, 130

Germany, With Army of Occupation in, 1918-19, 17-18
Goss, Leonard, 19
Government in Industry, F.M.'s views expressed in *Murphy News*, 108

Haden, W. J. (Swansea Murphy Dealer), 78
Hand, Norman (married Audrey Murphy in 1950), 182
Harrop, Cedric and Laurel, 178-179
Herod, Arthur (first Murphy Radio representative, later Area Manager), 35
Herod, Hugo (Northern Ireland representative), 98
Highgate Road Chapel, 11

Hills Bros. (Upton Park Murphy Dealer), 99
Hitler, Adolf, joke about, 143-144; against freedom, 126
Hood, Alfred (F.M.'s driver), 117
Howe, Charles and Edith (parents of Hilda Howe), 12-13
Howe, Hilda (later Murphy), 12-14, 17, 18, 20, 22, 24, 25, 32, 33, 49, 138, 168, 173, 174, 186, 187
Hum, Jack (Editorial department of *Murphy News*, 1933-55; Editor, 1955-56), Foreword, 91, 99, 190
Hungarian manufacturer, 174

Integrity, basic principle of, 5; definition of, 77, 129
International Correspondence Schools, post as examiner, 11

Jane (in the *Daily Mirror*), 69
Japanese competition in electronics, 123
Joan—see Murphy family

Kenneth—see Murphy family
Kent, E. W. (Distribution Manager, Murphy Radio), 77, 107
Knowles, Stanley, photographic re-toucher, 20
Knowlson, H. (Abergele Murphy Dealer), 89

Law, Equality before the, 132
Leggo, Annie (Murphy), 8-9
Light, Pat (Cussins & Light), 82-83
Limited Dealership Scheme, 3
Limited Liability Company, modifications to, in New Conception of Business, 127-129
Livingstone, Sir Richard, 109, 125
Ludwick Corner, Welwyn Garden City, 139-142, 168

Making Wireless Simple, 30, 35, 61
Man with the Pipe (F.M.'s nickname from his advertisements), 37, 39, 79, 121
Manton, R. E. (Clapham Murphy Dealer), 99
Marconiphone advertisement, 39; Model 55 portable, 52
Marshall, Esther (friend of Hilda Howe), 13
Maurice—see Murphy family
McIntosh of Kirkcaldy, 156
Milton, Canada, farm at, 178, 179
Mullard, (valves), 21
Murphy Casson Ltd. (formerly Engineering Publicity Service), 22, 24, 25, 34
Murphy Dealers, 73-92
Murphy family: John James and Annie (F.M.'s parents), 8-9
    Arthur, 14; Ethel, 14, 17, 117; Evelyn, 14; Harold, 14; Leonard, 14; Winifred, 14; (F.M.'s sisters and brothers)
Murphy, Kenneth (eldest child of F.M., born 5.6.13), 12; at school, 22; married Audrey Mary Phillips, worked at Murphy Radio, 105; joined Frank Murphy Ltd., 135; set up retailing organisation, 159; opened guest house, 168; war-

time ambulance service, 169; died, 174
Murphy, Joan (second child of F.M., born 4.6.18), at school, 22; at college, 169; marriage to John Long dissolved, 187; contributions to *Murphy Review*, 149-151, 164
Murphy, Maurice (third child of F.M., born 16.5.25), 22; poem in *Murphy Review*, 149; left school, 169; emigrated to Canada, 178; helped F.M. by making coffee tables, 179; joined R.C.A.F., 180; married Carol Maddock, 183; invited H.M. to Canada, 186; flown home from France to attend F.M.'s funeral and settle his affairs, 189
Murphy, Ann (daughter of Kenneth and Audrey Murphy), 186, 189
Murphy, Patrick (son of Kenneth and Audrey Murphy), 186, 189
See also the Murphy Family Tree—Appendix 1.
Murphy, Frank born 16.6.1889. Early life, up to end of World War I, 8—18; marriage to Hilda Howe, q.v., 12; meeting with C. R. Casson, 19; start of E.P.S., 20; reasons for starting Murphy Radio, 26; basic principles through life, 1-7; Limited Dealership scheme 3; dealer's discount cut to 27½%, 76; personally featured in M.R. advertising ("The Man with the Pipe"), 37, 39; attitude to Radiolympia, 48; to unemployment, 45-46; to profit-making, 52; Rent Theory, 85; trip to United States, 117; final Murphy Radio Board Meeting, 120; formulated New Conception of Business, 124-132; formed Frank Murphy Ltd. (furniture), 133; summoned to Law Courts, 169; jailed for contempt of court, 169; worked for Admiralty, 172; son Kenneth died, 174; started Frank Murphy of London Ltd. (timber products, radio), 176; started One World Laboratories, 177; emigrated to Canada, 178; divorced by Hilda Murphy, 175; married Audrey Murphy, but marriage later annulled, 180; various small-time jobs in Toronto, 181-182; designed radio-phonograph, 183; formed Murphy Partnership, 185; asked Hilda to re-marry him, 187; mathematics teacher at Bloor Collegiate School, 189; died from heart attack, 189
Murphy Madness (dealers' enthusiasm), 93 and 73-92
Murphy News (House organ of Murphy Radio, became forum for free speech under editor Stan Willby), 93-123
Murphy Partnership (unsuccessful venture in Canada), 185
Murphy Radio Ltd.. company formed in 1929, 27; early research, 29-31; first set sold, 110; progress in first six years, 52; first superhet receiver (A8), 53; first console 55; first radio-gramophone, 56; first television set, 58; take-over by Rank Organisation, 122; "Murphy" trade name acquired by J. J. Silber, 123
Murphy Review (monthly magazine of Frank Murphy Ltd.), 133-167
Mosley, Sir Oswald, Fascism in Britain, 114; interned in Brixton prison under "18B" regulations, 171

Naz, J. G. 112. 139, 147
New Conception of Business, 124-132
Nichol, James (Maitland Bros., Edinburgh Murphy Radio Dealers), 89
Noise suppression, device for in Murphy sets, 56

O'Brien, P. K., (Murphy Radio economist), 77, 115, 122
O'Brien, T. H., (Frank Murphy Ltd. accountant), 164, 165

Officers' Wireless School, Farnborough, in World War I, job of starting given to Frank Murphy, 17
Owen, T. L. (Frank Murphy of London Ltd.), 160
Oxford University, Frank Murphy awarded a mathematical exhibition to study at, but had to abandon course, 10

Palmer, Fred (builder in Welwyn Garden City), 168
Pank, John (Norwich Murphy Radio Dealer), 145
Parratt, D. W. (Head of Service Department at Murphy Radio), 83
Partnership, Murphy, 185
Patent Office, 10
Paterson, J. (Frank Murphy of London Ltd.), 160
Perry, Ronald (Frank Murphy Ltd.), formed company with Boyd. 166
Philips Electronics, 187-188
Post Office (Engineering Research Section), 10-11
Power, Edward J. (Chief Engineer, Murphy Radio, 1929-37, Managing Director, 1937-62), 27, 57, 59, 63, 76, 77, 111, 119, 121
Price of Murphy Sets, how calculated, 52
Problem-solving, one of Frank Murphy's basic tenets, 6
Proctor, G. C. (Eastern Counties representative for Murphy Radio), 98
Publicity Club of London 86

Questionnaires, used in research, 148, 154, 155, 156; difficulty of interpreting results, (Boyd), 157

Radiogramophone, Murphy Radio's first, 56
Radiolympia, 48, 53
Rank Organisation, 123
Rank Radio International, 123
Rank Toshiba, 123
Rent Theory, 85, 122
Reproduction, perfect, said to be unobtainable, 31
Retailing, theories of, 86, 87, 107
Ridley, J. B., 160, 176
Robertson, J. Craig, 160, 176
Rosen, L. (Croydon Murphy Dealer), 98
Rowney, Tom (Stafford Murphy Dealer), 82
Royal Flying Corps, 15-18
Rule of Law, 126, 132, 175, 177, 185, 192
Rupke (Canadian entrepreneur), 187
Russell, Sir Gordon, C.B.E., M.C., asked to design cabinets for Murphy sets, 59; extract from his autobiography, "Designer's Trade", 59-72, 77, 101
Russell, Professor R. D. (Dick), 59, 63, 66, 67

Sales of Murphy sets, 1930-1935, 52
Science Museum, Murphy television donated to, 58
Seccombe, R. O., succeeded D. W. Parratt as Head of Service Dept., 83
Service to the Community, basic principle, 3

Shelton, Jim (Frank Murphy of London Ltd.), 176
Shop, What is a, 86, 87
Short-wave sets, 106
Showrooms for Murphy furniture, 158
Sideboard, Murphy, 153
Silber, J. J., acquired Murphy trade name, 123
Sims, A. J. (cabinet-maker, Frank Murphy Ltd.), 141
Stanley, C. O., 22
Station names in alphabetical order, 58
Stephenson, Clifford (Huddersfield Murphy Radio Dealer), 76, 77, 79, 99
Stocking Plan, 118
Superheterodyne receivers, Foreword, 53, 90
Swain, F. A. (Merriotts, Bristol Murphy Dealer), 119

Table, design of fixed, and design of drawleaf, 151-153
Television set, first Murphy Radio, 58
Thomas, Eric R. (Editor, *Murphy Review*), 141
Thompson, Frank (South Shields Murphy Dealer), 84
Toshiba, joint venture with Rank Radio to produce television sets, 123; later, set up own factory in Plymouth to manufacture under Toshiba name, 123
Trinder, Mrs. E. (Saltburn Murphy Dealer), 76, 91
Trustees of New Conception of Business, proposal for, 128
Tuning, special self-tuning device, 56, 90
Turner, Arthur Robinson (married Winifred Murphy); in Navy in World War I, 15; anti-pacifism, 113; Trustee of Frank Murphy Ltd., 127, 133
Turner, Laurence Beddome (Trustee of Frank Murphy Ltd.), 134

University of London, degree in electrical engineering awarded to Frank Murphy, 10
University of Oxford, scholarship to, not taken up, 10

Vallance, Alec (Mansfield "Murphy Mad" Dealer), 82, 89, 98
Value for Money, basic principle, 2, 52, 57, 80
Varnish, protective, for Murphy furniture, 145

Watch, gold, presented to Frank Murphy by Newcastle Dealers' Association, 92, 190
Wavelengths, calibration in, for the first time, 30, 37
Welwyn Garden City, 28, 32-33. 83
Western Electric Company, 10
Wilde, Mr. (Ilford Murphy Dealer), 77
Willby, Stanley (First Editor of *Murphy News*), 94, 99, 102, 108, 121
Williams, Prof. R. C. G. (Head of Research Team), 53, 122
Windsor chair, 155
Wireless School (R.F.C.) establishment of, 17
Wireless World, The, 90
Wood, Herbert G. (Trustee of Frank Murphy Ltd.), 134
Wood, W. C. (Cabinet-maker, Frank Murphy Ltd.), 141

World War I, Frank Murphy in, 15-18
World War II, effect of, on Frank Murphy Ltd., 163, 165, 166
Worley, E. W. (Worley & Sewter, Gateshead Murphy Dealer), 80-81

York & Sons (Kettering Murphy Dealer), 81

Fig 52. The Murphy Phonograph made in Canada (1948)

Fig 51. Frank with Maurice in Toronto (1953)

Fig 53. Hilda Murphy aged 85 with great-grand daughter Deborah Long (1972

Fig 54. Hilda Murphy aged 90 receiving a Murphy TV set

Fig 55. Last portrait of Frank Murphy, taken in Canada

O${}_{NE}$ asked me when the earth of my vision
   would come into being,
I answered :
In my heart it is already here ;
And though I cannot know when it will awaken
   in the hearts of others ;
Yet, the tension of life is so strong,
Uncertainty and suffering so widespread,
And the first notes of understanding so oft repeated,
That I feel the time is near

From ' Earth ' by Frank Townshend

Fig 56. Poem by Frank Townshend